Contents

1 Chapter 1: Introduction to Wind Energy

Wind energy harnesses nature's power, converting the kinetic energy of moving air into a usable form of electricity. This renewable energy source offers a sustainable alternative to fossil fuels, mitigating the harmful effects of greenhouse gas emissions and contributing to a cleaner, healthier environment. Its increasing popularity stems from its accessibility and potential for large-scale deployment. Understanding the fundamentals of wind energy is crucial for anyone interested in sustainable energy solutions.

The availability of wind energy is largely dictated by geographical location and prevailing weather patterns. Areas with consistent, high-speed winds are ideal for wind farm development. Coastal regions and elevated terrains often experience stronger and more reliable winds compared to flat inland areas. However, wind speed is notoriously variable, influenced by topography, vegetation, and even time of day. This variability necessitates sophisticated energy storage solutions and grid management strategies to ensure consistent power delivery. Understanding these variations is key to effective wind farm planning and operation.

Harnessing wind energy relies on specialized structures known as wind turbines. These are essentially sophisticated windmills, designed to capture the energy from the wind and convert it into rotational motion. This rotation then drives an electrical generator, producing electricity that can be fed into the power grid. Modern wind turbines are incredibly efficient, capable of extracting a significant amount of energy from even moderate winds. Their design incorporates advanced materials and aerodynamic principles to maximize energy capture and minimize energy loss. These technological advancements have resulted in significantly improved efficiency and reliability of wind energy generation.

Several factors must be considered when assessing the viability of a wind energy project. These include the wind resource at the potential site, the capacity of the wind turbines, the proximity to transmission infrastructure, and land availability. Environmental considerations, such as the impact on local ecosystems, visual impact, and noise pollution, must also be carefully evaluated. Furthermore, regulatory processes, permitting requirements, and community engagement play critical roles in project success. A comprehensive approach, involving thorough planning and consideration of all relevant factors, is necessary for responsible and sustainable wind energy deployment.

Wind energy has a rich history, evolving from basic windmills used for centuries for mechanical work to the highly sophisticated turbines we see today. Early windmills were primarily used for pumping water and grinding grains, representing an early form of harnessing wind energy for practical purposes. The development of electricity brought about a major shift, enabling the conversion of wind energy into electricity, creating a more readily usable form of energy. Technological advancements over the past few decades have substantially increased wind turbine efficiency, making wind power a more economically viable and environmentally friendly energy source.

Technological advancements are pushing the boundaries of wind energy. The design of wind turbine blades is constantly improving, incorporating more advanced materials and aerodynamics. This leads to higher energy capture rates and increased efficiency, optimizing turbine performance. Improvements in generator technology result in higher power output and increased reliability. Sophisticated control systems enhance efficiency by automatically adjusting turbine operation based on wind speed and direction. These continuous innovations reflect the ongoing commitment to improving the effectiveness and sustainability of wind energy.

The integration of wind energy into existing power grids presents both opportunities and challenges. The intermittent nature of wind power, varying with wind speed and direction, requires effective grid management strategies to ensure power stability. Smart grid technologies play an essential role in addressing this intermittency, enabling better prediction, integration, and control of wind power generation. These technologies allow for better management of fluctuating electricity supply and demand. Their development is crucial for the wider adoption of renewable energy sources like wind power. The seamless integration of wind farms into existing infrastructure is vital for maximizing their contribution to a sustainable energy future.

The economic aspects of wind energy involve initial investment costs, operational expenses, and potential revenue streams. Government incentives and policies play a crucial role in shaping the economic viability of wind power projects. Subsidies, tax credits, and renewable portfolio standards can significantly reduce the cost of wind energy and accelerate its adoption. The increasing competitiveness of wind power against conventional energy sources and the long-term benefits of clean energy further contribute to its economic attractiveness. Thorough cost-benefit analyses are vital for making informed decisions about wind energy investments.

Successful implementation of wind energy projects requires careful planning and meticulous execution. Site selection is crucial, balancing factors such as wind resource availability, environmental impact, and proximity to the grid. Environmental impact assessments are critical in identifying and mitigating potential negative consequences. Proper permitting and regulatory compliance are essential for legal and sustainable operation. The successful wind farms around the world serve as powerful examples of what can be accomplished with strategic planning and rigorous implementation. By learning from past successes and failures, we can further enhance the efficiency and effectiveness of wind energy projects.

Offshore wind energy represents a significant untapped potential. The consistent and higher wind speeds available offshore can lead to higher energy generation capacity. However, the construction and maintenance of offshore wind farms present unique challenges, including harsher environmental conditions and higher costs. Nonetheless, the substantial potential of offshore wind power makes it a key area for future development. Continued technological advancements are essential to mitigate the challenges associated with offshore wind energy, unlocking its vast potential as a clean and sustainable energy source.

1.1 What is Wind Energy?

Wind energy, simply put, harnesses the kinetic energy of moving air—the wind—to generate electricity. It's a clean, renewable resource, unlike fossil fuels which are finite and contribute significantly to climate change. The wind's power, however, is inconsistent; its speed and direction vary constantly, posing challenges in terms of consistent energy production. This variability is a key aspect to understand when considering wind energy's role in the broader energy mix.

The process of converting wind energy into usable electricity begins with wind turbines. These towering structures, often seen dotting landscapes, use rotating blades to capture the wind's kinetic energy. As the wind pushes against the blades, they spin, driving a generator within the turbine. This generator, in essence, is a complex system of magnets and coils of wire, converting mechanical rotation into electrical current. The electricity generated then flows through transmission lines, ultimately reaching homes and businesses.

Understanding the physics involved is crucial. The amount of power a wind turbine can generate is directly related to the cube of the wind speed. This means a doubling of wind speed results in an eightfold increase in power output. However, extremely high wind speeds can also damage the turbine, necessitating sophisticated control systems to regulate blade pitch and prevent over-speeding. These systems continuously monitor wind conditions, adjusting the turbine's operation for optimal performance and safety. The design of the turbine blades themselves is a critical factor, influencing their efficiency in capturing the wind's energy and converting it into rotational force. Years of research and development have yielded highly optimized blade shapes and materials. The location of a wind turbine—or more accurately, a wind farm—is paramount to its effectiveness. Wind resource assessment is a significant undertaking, involving detailed meteorological studies to determine the average wind speed, turbulence intensity, and wind direction at potential sites. Factors such as terrain, elevation, and proximity to obstacles all play a role. Choosing the right location significantly impacts the overall cost-effectiveness of the project, as a site with consistent high-speed winds will yield greater energy production. Poor site selection can lead to underperformance and financial losses.

Beyond energy production, the environmental impact of wind energy deserves careful consideration. While wind power is significantly cleaner than fossil fuels in terms of greenhouse gas emissions, there are environmental concerns to address. These include potential impacts on wildlife, particularly birds and bats, and the visual impact of large wind turbines on the landscape. Careful siting practices, using environmental impact assessments and mitigation strategies, can minimize these negative effects. Moreover, the manufacturing and disposal of wind turbine components, such as blades and gearboxes, requires consideration in a complete lifecycle analysis. These materials require responsible management to prevent negative environmental consequences.

Finally, the economic aspects are equally important. The initial investment costs for wind farms are considerable, encompassing the purchase of turbines, construction, land acquisition, and grid connection. However, once operational, wind energy has relatively low operating costs, primarily involving routine maintenance. Government policies, such as tax credits, subsidies, and renewable energy mandates, play a significant role in making wind energy economically viable. These policies can reduce the upfront financial burden and stimulate investment, driving the growth of wind power as a significant part of our energy supply. Furthermore, the economic impact extends beyond the energy sector, creating jobs in manufacturing, installation, maintenance, and research.

1.2 Wind Energy Resources

Understanding wind resource availability is crucial for successful wind energy projects. We need to move beyond simply knowing that wind exists; we must quantify its potential. This involves detailed analysis of wind speed, direction, and consistency over extended periods. Sophisticated instruments, such as anemometers and wind lidar, are used to gather this data. The data collected forms the foundation of a comprehensive wind resource assessment, crucial for determining the economic viability and environmental impact of a project. This assessment must include the analysis of wind shear, wind turbulence and other relevant factors.

Accurate wind speed data is essential. Think of it as the lifeblood of the entire project. Different locations will naturally have varying wind speeds and patterns, dictated by geographical features like hills and coastlines, and even prevailing weather systems. Analyzing this data helps us pinpoint the locations that offer the most consistent and powerful winds, thereby maximizing energy production. The longer the period of data collection the better, as this will help minimize the uncertainties linked to the variability of weather patterns.

Locating ideal sites often requires a multi-faceted approach. Topographical surveys are needed to identify potential areas with consistent high wind speeds, while environmental impact assessments are crucial to ensure minimal disruption to ecosystems and wildlife. These assessments aren't simply checklists; they require careful consideration of noise pollution, visual impact, and potential effects on local bird populations. Regulations and permitting processes can be complex and vary by location, adding another layer of consideration.

Beyond just land-based wind resources, there's a burgeoning world of offshore wind. Offshore sites typically boast more consistent and higher wind speeds compared to onshore locations. However, the challenges are greater. Constructing and maintaining offshore turbines requires specialized equipment and techniques, and the additional costs must be carefully considered. The environmental impact assessments for offshore projects are also particularly complex, encompassing potential effects on marine ecosystems, including fish populations, sea mammals and other living creatures.

The variability of wind is a significant factor. Wind resources are inherently unpredictable, unlike fossil fuels. This intermittency poses challenges for grid integration, demanding innovative solutions to maintain stable power supplies. Energy storage technologies, such as pumped hydro storage and batteries, are crucial for managing this variability. They allow us to store excess energy during periods of high wind and release it when the wind is low. Smart grid technologies play a pivotal role in optimizing the flow of energy from wind farms, ensuring a balanced supply and demand even with fluctuating wind speeds.

Predicting wind power output requires sophisticated modeling techniques. These models take into account various factors, including wind speed, turbine specifications, and atmospheric conditions. Advanced meteorological data and complex algorithms are used to estimate the potential energy output of a wind farm over a given period. These predictions are critical for economic feasibility studies and investment decisions. Accurate forecasts allow for better planning and management of energy distribution, minimizing the reliance on backup power sources.

Furthermore, understanding the long-term potential of a wind resource requires a careful assessment of climate change impacts. Changes in wind patterns and precipitation could alter the productivity of existing wind farms. Therefore, comprehensive assessments must consider the potential implications of long-term climate patterns on wind power output. Adaptive strategies to address these changes are also vital in long term resource planning. It's important to recognize that wind energy, while clean, isn't without its own environmental considerations. The manufacturing process of wind turbines involves the use of materials and resources. Careful consideration of these factors is essential during the project lifecycle, from inception to decommissioning. Minimizing environmental impact requires a holistic and integrated approach throughout the project timeline.

Finally, remember the importance of community engagement. The successful implementation of wind projects requires the support and understanding of local communities. Transparent communication and meaningful consultations can address concerns regarding visual impact, noise levels, and potential disruption to the environment. This open dialog can greatly enhance acceptance and promote positive community relations. These steps are crucial for long-term successful wind energy generation. Effective community relations can translate directly into successful projects. Ignoring this aspect can lead to delays, opposition, and ultimately, project failure.

2 Chapter 2: Harnessing the Wind

Harnessing the wind's power isn't a new concept; humans have been using wind energy for millennia, initially with simple windmills for grinding grain and pumping water. These early inventions, while rudimentary, laid the groundwork for the sophisticated wind turbines we see today, demonstrating the enduring appeal and practicality of wind power as a renewable energy source. Consider the dramatic shift from these basic mechanisms to the advanced engineering marvel that is a modern wind turbine. This evolution highlights the remarkable progress made in harnessing wind's energy, transforming it into a significant contributor to our global energy mix.

The fundamental principle behind wind energy conversion is remarkably straightforward: wind's kinetic energy – its motion – is captured by the blades of a wind turbine. These blades, often made from composite materials, are designed to efficiently convert wind's kinetic energy into rotational energy. This rotation, in turn, drives a generator, a device that converts mechanical energy into electricity. The system is elegantly simple, but the underlying engineering and precision are sophisticated. It's not merely about placing a turbine in a windy spot; optimizing turbine design and placement requires a deep understanding of aerodynamics, material science, and electrical engineering.

Several types of wind turbines exist, each with unique characteristics and applications. Horizontal-axis wind turbines (HAWTs), the most common type, resemble the large wind turbines seen in wind farms across the globe. Their horizontal axis of rotation allows for efficient capture of wind energy, particularly at higher wind speeds. Vertical-axis wind turbines (VAWTs), on the other hand, are characterized by a vertical axis of rotation. While less common than HAWTs, VAWTs offer advantages in certain situations, such as the ability to capture wind from multiple directions. This makes them potentially suitable for urban environments or locations with less predictable wind patterns. The choice of turbine type depends heavily on factors such as wind resource characteristics, land availability, and economic considerations.

Beyond the basic design, turbine technology encompasses various components that collectively determine efficiency and performance. Blade design is crucial, as it dictates how effectively the wind's energy is captured and converted. Aerodynamic principles govern the shape and curvature of the blades, maximizing lift and minimizing drag. Similarly, the gearbox plays a vital role in transferring the relatively slow rotational speed of the turbine blades to the higher rotational speed required by the generator. Advanced generators, often using permanent magnets, convert this mechanical energy into electricity efficiently and reliably. Control systems regulate the turbine's operation, adjusting blade pitch and yaw to optimize energy capture and protect the turbine from extreme weather conditions. These systems, coupled with sophisticated monitoring, ensure both efficient operation and longevity of the turbine system.

The tower, a critical structural component, elevates the turbine nacelle and rotor to capture the higher, faster wind speeds that are generally found at greater heights. The selection of the tower's design is influenced by factors such as site conditions and wind resource characteristics. Materials and designs are selected based on structural integrity, cost-effectiveness, and ease of transportation and assembly. Furthermore, regular maintenance and repair are essential to ensure the long-term reliability and operational efficiency of the entire wind turbine system. Preventive maintenance schedules, optimized lubrication, and prompt response to any identified issues are critical for extending the operational lifespan and maintaining high power output of wind turbines. The economic implications of proper maintenance cannot be overstated, as unplanned downtime can significantly impact energy production and financial returns.

Ultimately, the successful harnessing of wind power depends on the synergistic interaction of these various components and design choices. Optimizing turbine placement, understanding the intricate interplay of aerodynamic principles, selecting appropriate materials, implementing efficient control systems, and diligently carrying out routine maintenance are all crucial aspects of successfully extracting renewable energy from the wind. Only through a holistic approach to design, implementation, and operation can the full potential of wind energy be truly realized. Understanding this interconnectedness is essential for both designing and operating wind energy systems effectively and sustainably. The long-term sustainability and viability of any wind farm requires not just technological mastery but a keen understanding of the environment and responsible energy generation practices.

2.1 History of Wind Power

Harnessing wind's power isn't a recent invention; its history stretches back millennia. Early civilizations, recognizing the wind's potential, ingeniously devised ways to utilize its force. Ancient Persia, for example, employed windmills for grinding grain as early as the 7th century, a testament to humanity's early understanding of harnessing natural energy. These were primarily vertical-axis windmills, their design reflecting the available technology and the specific needs of their time. Their simple yet effective mechanisms allowed them to perform essential tasks, profoundly impacting daily life in agricultural communities across vast regions. The knowledge of wind power, though rudimentary by modern standards, spread gradually across continents, adapted and refined by different cultures to suit their environments and purposes. The impact of these early windmills was significant, underscoring the enduring human drive to tap into renewable resources. Their legacy continues to inform modern designs, demonstrating the enduring relevance of centuries-old ingenuity.

The widespread adoption of windmills in Europe, particularly during the Middle Ages, marked a significant turning point. Their use expanded beyond simple grain milling to encompass tasks like pumping water and sawing wood. Dutch windmills, renowned for their elegant design and efficiency, became iconic symbols of technological advancement. These sophisticated structures, with their intricate gear systems and adjustable sails, represented a significant leap in wind energy utilization. Their construction often involved elaborate engineering solutions, demonstrating a remarkable level of craftsmanship and an advanced understanding of mechanical principles for the era. They played a crucial role in land reclamation, water management, and industrial processes, showcasing the versatility and practicality of this sustainable energy source. The impact of Dutch windmills extended far beyond the Netherlands, influencing windmill design and implementation in other parts of Europe and beyond.

The 19th century saw significant technological advancements, paving the way for a new era of wind power. Improvements in materials science and engineering led to stronger, more reliable windmills, capable of generating greater power. The advent of electricity brought about another profound shift, transforming windmills from purely mechanical devices to generators of electrical power. This was a critical step towards the integration of wind power into modern energy systems. Early attempts to generate electricity using wind were often experimental, but they proved the feasibility of this approach and laid the groundwork for future developments. The development of increasingly efficient generators allowed wind turbines to deliver larger amounts of electricity, proving critical for increasing power output. These pioneering efforts, though often hampered by technological limitations, held immense potential and demonstrated the increasing potential of this form of energy generation.

The 20th century witnessed an explosion in wind turbine technology. The advent of more efficient blades, improved materials, and advanced control systems led to the development of larger, more powerful wind turbines. The latter half of the century saw the rise of modern wind farms, showcasing the potential of wind power as a viable source of energy on a large scale. This period also saw the development of both horizontal-axis and vertical-axis wind turbines, reflecting innovation in different aspects of turbine design and their suitability for diverse environments. Advances in materials science led to the production of stronger, lighter, and more aerodynamic blades, capable of extracting more energy from the wind. These developments not only increased efficiency but also enabled the construction of larger turbines, thus further maximizing energy output. The shift toward large-scale wind farms significantly increased the role of wind power in global energy production.

The late 20th and early 21st centuries have seen an extraordinary acceleration in wind energy's growth. Technological advancements, coupled with increased environmental awareness and government support, have propelled wind power to become a major contributor to global renewable energy supplies. The cost of wind power has significantly decreased, making it economically competitive with traditional energy sources. Furthermore, advancements in offshore wind turbine technology have unlocked enormous untapped resources. This progress shows the potential of wind energy to play an even larger role in a sustainable energy future. Continued innovation in blade design, energy storage, and grid integration technologies continues to expand its potential and improve its efficiency, shaping the development of wind power for decades to come. The story of wind power is one of continuous innovation and adaptation. It is a story of recognizing the power of nature and harnessing it for human benefit, transforming centuries-old ingenuity into a major source of sustainable energy.

2.2 Wind Turbine Basics

Wind turbines, at their core, are elegant machines converting wind's kinetic energy into usable electricity. This process begins with the wind itself, a natural resource constantly replenished. Understanding this fundamental principle is crucial for comprehending how a wind turbine functions effectively. The wind interacts with the turbine blades, causing them to rotate. This rotational energy is then transferred through a sophisticated system of components.

The rotor, comprising two or more blades, acts as the primary energy collector. These blades, often made from composite materials, are carefully designed with specific aerodynamic profiles to maximize the amount of energy extracted from the wind. The shape and length of the blades are optimized to capture the maximum energy from a wide range of wind speeds. This is a critical aspect of turbine efficiency and overall power generation. Manufacturers invest heavily in research and development, constantly striving for improvements in blade design.

From the rotor, the rotational force travels to a gearbox. The gearbox increases the rotational speed, transferring it to the generator. This is a crucial step because the generator requires a high rotational speed to generate electricity efficiently. The gearbox, a complex mechanism, must withstand immense stress and is a key maintenance focus for wind turbine operators. Understanding gearbox function is vital for comprehending wind turbine maintenance needs and optimizing operational efficiency.

The generator, housed within the nacelle (the casing atop the tower), converts mechanical energy into electrical energy using electromagnetic induction. This process involves rotating magnets within coils of wire, producing an electrical current. The electrical current generated is then passed through a rectifier, converting it into direct current (DC). The DC power is then converted to alternating current (AC) through an inverter, suitable for transmission into the electrical grid. The efficiency of the generator, therefore, directly impacts the amount of electricity a turbine can produce from the wind.

The tower itself supports the entire wind turbine assembly, elevating the rotor to capture stronger and more consistent winds. Tower height significantly influences the amount of energy the turbine can generate. Taller towers place the rotor in higher wind speeds, resulting in a greater energy output. However, taller towers also require more substantial construction and present greater challenges in maintenance and repair. The tower's structural integrity is paramount for the turbine's operational safety and longevity. Choosing the correct tower height and design for a particular site involves careful consideration of several factors such as wind speed at different heights, terrain features, and construction costs.

Control systems are also essential for safe and efficient wind turbine operation. These systems monitor wind speed, yaw (the turbine's orientation to the wind), and other parameters. If wind speeds exceed safe limits, the control system will automatically stop the turbine to protect it from damage. These systems also manage the pitch of the blades, optimizing energy capture and reducing wear and tear. Advanced control systems are a key feature of modern wind turbines, improving performance and reliability while minimizing unnecessary stress on the components. Effective and efficient control systems are crucial to maintain the long-term functionality of a wind turbine.

Finally, the integration of wind turbines into the electrical grid requires careful consideration. The intermittent nature of wind energy requires smart grid technologies to manage power fluctuations and ensure grid stability. These technologies play a critical role in ensuring that the power generated by wind turbines is reliably integrated into the electricity supply. Understanding how these technologies function is pivotal for building a sustainable energy future. This involves sophisticated forecasting models and energy storage solutions that can smooth out the intermittent output from wind farms.

The overall efficiency of a wind turbine depends on the interaction of all these components and the effectiveness of their integration. Understanding each component's role and how they work together is paramount to maximizing the energy output of a wind turbine and to ensuring its efficient and safe operation. Regular maintenance and monitoring are also essential for extending the lifespan and optimizing the performance of the wind turbine, ensuring a long-term return on investment. Regular checks on the gearbox, the generator, and other critical components will minimize the likelihood of malfunctions. A well-maintained wind turbine not only guarantees uninterrupted electricity production but also enhances safety and extends the operational life of the equipment.

2.3 Types of Wind Turbines

Horizontal-axis wind turbines (HAWTs) dominate the current wind energy landscape. Their design, with a horizontal rotor mounted on a tower, is familiar, even iconic. This configuration efficiently captures wind energy due to the streamlined alignment of blades with prevailing winds. The majority of HAWTs utilize three blades, a balance between maximizing energy capture and minimizing structural complexity and cost. However, two-bladed designs exist, particularly in older installations or niche applications, although they present greater challenges in terms of vibration control and structural fatigue. The size of HAWTs varies considerably, ranging from small, domestic-scale turbines to massive offshore behemoths with rotor diameters exceeding 200 meters. These colossal turbines can generate megawatts of power, significantly contributing to national power grids. Choosing between two or three blades involves careful consideration of factors like wind conditions and cost-effectiveness, a decision often heavily influenced by site-specific analyses.

Upwind HAWTs position the rotor upstream of the tower, offering advantages in terms of aerodynamic efficiency and reduced tower shadow effect, optimizing the airflow to the blades. Downwind HAWTs, on the other hand, place the rotor downwind of the tower. While this design is typically simpler and less expensive to manufacture, it poses additional challenges, such as increased dynamic loads on the tower and the need for sophisticated yaw control systems to ensure the turbine stays aligned with the wind.

The choice between upwind and downwind designs is a complex engineering trade-off, influenced by the prevailing wind regime, site constraints, and the desired level of efficiency and operational reliability. Ultimately, both configurations play a role in the global wind energy mix, each best-suited for different conditions and applications.

Vertical-axis wind turbines (VAWTs), in contrast to their horizontal counterparts, feature a vertical rotor axis. This configuration offers several intriguing advantages. Their rotational axis is independent of wind direction, eliminating the need for complex yaw mechanisms, simplifying the overall design and reducing operational costs. This design characteristic proves particularly beneficial in locations with highly variable or unpredictable wind patterns. Moreover, VAWTs are typically shorter and more compact than HAWTs, sometimes making them more suitable for urban environments or sites with space limitations. They can also offer quieter operation due to the reduced noise generated by the rotating components.

However, VAWTs also present distinct challenges. Their aerodynamic performance is generally less efficient than HAWTs, meaning they tend to generate less power for the same wind resource. This lower efficiency stems from the complex flow patterns around the rotor, which make maximizing energy extraction significantly more difficult than with HAWTs. In addition, the torque variations in VAWTs can be greater, which necessitates robust drivetrain components to handle the fluctuating loads. Further research and development are crucial to overcome these efficiency hurdles and to improve the overall competitiveness of VAWTs in the wider energy market. Consequently, they still represent a smaller, albeit growing, segment of the wind energy sector.

Darrieus turbines, a common type of VAWT, are characterized by their curved blades that rotate around a vertical axis. Their aerodynamic performance is critically dependent on the blade shape and their interaction with the surrounding airflow. Optimizing their blade design is complex, demanding sophisticated computational fluid dynamics simulations and extensive experimental testing to maximize power extraction. The design's unique shape also leads to complex stress patterns in the blades, demanding high-strength materials and careful structural analysis. Careful consideration needs to be given to the materials chosen for their production, ensuring both durability and efficiency to maximize the turbine's lifespan.

Savonius turbines are another notable type of VAWT, distinguished by their simple, S-shaped blades. These rotors are inherently less efficient than Darrieus turbines, but their straightforward design and robustness make them suitable for applications where simplicity and low cost are prioritized over high power output. Their simple construction makes them easier to manufacture and maintain, a clear advantage in regions with limited infrastructure or expertise. Their applications frequently involve low-power tasks, like water pumping in remote areas, where simplicity and durability are key considerations that trump the need for high efficiency. While their energy generation capacity is lower compared to other turbine types, their cost-effectiveness and simplicity make them valuable contributors in specific applications, enhancing access to energy in less accessible areas.

In summary, the choice of wind turbine—HAWT or VAWT, and their specific subtypes—depends heavily on multiple factors. These include the site's specific characteristics like wind regime, space constraints, and grid connection possibilities. Economic factors such as installation costs, maintenance requirements, and anticipated energy yield all weigh heavily in the decision-making process. Understanding these different designs, their advantages, and limitations empowers engineers and planners to make informed decisions when selecting the most suitable turbines for maximizing sustainable energy generation at various scales, ranging from individual households to massive utility-scale wind farms. The future of wind energy relies on continual innovation and improvement in both HAWT and VAWT technologies, each with the potential to play a crucial role in shaping a sustainable energy future.

3 Chapter 3: Turbine Technology

The heart of any wind turbine lies in its intricate components, working in concert to capture the wind's energy and convert it into usable electricity. Understanding these components is crucial for appreciating the technology's sophistication and potential. We'll begin with the blades, the turbine's first line of contact with the wind, whose design is a masterful blend of aerodynamics and material science. The airfoil shape of each blade, meticulously engineered, is crucial for maximizing lift and minimizing drag. Modern designs incorporate advanced materials like fiberglass and carbon fiber composites, enabling the creation of lighter, stronger, and more efficient blades that withstand the stresses of high wind speeds and harsh weather conditions. Consider the challenges of blade design: achieving the optimal balance between weight, strength, and aerodynamic performance; designing blades to operate efficiently across a wide range of wind speeds; and developing robust blades capable of withstanding extreme weather events. These are not simple problems. These considerations highlight the complexities of blade design and the continuous effort to improve their performance.

Next, we turn to the gearbox, a critical component that manages the rotational speed differences between the slow-turning rotor and the high-speed generator. This crucial translation of speed is needed to generate electricity at a frequency compatible with the power grid. Gearboxes are highly engineered mechanical systems, designed to withstand enormous torque and operate reliably under constant stress. Different gear designs, including planetary and helical gearboxes, each have their advantages and disadvantages in terms of efficiency, weight, and noise production. Recent technological developments explore gearless designs, directly connecting the rotor to the generator, which offers several potential benefits, especially in terms of increased efficiency and reduced maintenance. These innovative approaches reduce reliance on traditional mechanical components, leading to increased overall system reliability.

The generator, which converts mechanical energy into electrical energy, is another pivotal part of the turbine. Many modern wind turbines use permanent magnet synchronous generators (PMSGs), known for their high efficiency and compact design. Other types include doubly-fed induction generators (DFIGs), which offer variable speed capabilities, allowing the turbine to capture energy from a wider range of wind speeds. The choice of generator type influences the overall performance and cost-effectiveness of the turbine. Advances in generator technology continuously improve efficiency and power output, leading to more electricity being produced from each wind turbine.

Supporting the entire system is the tower structure, often made of steel or concrete. Tower height significantly impacts turbine energy production, as higher towers expose the blades to stronger, more consistent winds. The design of the tower must account for wind loads, seismic activity, and the overall weight of the turbine. Structural integrity is paramount; a failed tower can have devastating consequences. Consider the environmental impact of tower construction; selecting sustainable materials and minimizing the environmental footprint of the manufacturing process are important considerations. Innovation in tower materials and construction techniques is ongoing, to improve the efficiency and sustainability of wind turbine towers.

Finally, maintaining and repairing a wind turbine is an essential aspect of ensuring its long-term operation and economic viability. Regular inspections, preventative maintenance, and timely repairs are crucial for preventing costly downtime and maximizing energy production. Specialized technicians, with extensive training and safety protocols, perform these tasks. Predictive maintenance techniques, using sensors and data analytics, help in identifying potential problems before they escalate into major failures. Minimizing downtime through effective maintenance programs is economically vital, maximizing the lifespan and energy output of these large and expensive assets. The cost and complexity of maintenance underline the importance of meticulous planning and preventive care, thus maximizing the long-term effectiveness of wind turbines.

3.1 Blade Design & Aerodynamics

The efficiency of a wind turbine hinges significantly on its blades. These aren't simply spinning propellers; they're meticulously engineered aerodynamic surfaces, designed to extract maximum energy from the wind. Understanding blade design, therefore, is paramount to comprehending wind turbine technology. The shape, material composition, and overall design directly influence the amount of power a turbine generates.

Blade design starts with the airfoil profile. This is the cross-sectional shape of the blade, and it's crucial for lift generation. Similar to an airplane wing, the asymmetrical airfoil creates a pressure difference between the upper and lower surfaces when wind flows over it. This pressure difference generates lift, causing the blade to rotate. The precise curvature and thickness of the airfoil are optimized through computational fluid dynamics (CFD) simulations, allowing engineers to fine-tune performance for specific wind conditions. Variations in airfoil profiles along the blade's length (from root to tip) are common, adjusting lift and drag characteristics to maximize energy capture across the entire blade.

Material selection is equally critical. Early blades were often made of wood or fiberglass, but modern designs frequently utilize composite materials such as fiberglass-reinforced polymers or carbon fiber. These materials offer a high strength-to-weight ratio, allowing for longer, lighter blades that can withstand high rotational speeds and strong wind gusts. The weight reduction is crucial for minimizing stress on the turbine's gearbox and bearings, prolonging their lifespan and improving efficiency. Additionally, the stiffness and fatigue resistance of these materials are essential for ensuring the structural integrity of the blades over their operational life.

Beyond the airfoil and material choices, blade geometry significantly affects performance. The twist of the blade, which varies along its length, is a key design parameter. The twist is designed to optimize the angle of attack of each section of the blade, ensuring that each part of the blade operates at its most efficient lift-to-drag ratio across a range of wind speeds. The pitch of the blade, the angle between the blade and the plane of rotation, can be actively adjusted to optimize energy capture at different wind speeds. This active pitch control is a crucial feature in modern wind turbines, allowing for efficient operation under varying wind conditions and preventing damage during high-wind events.

Aerodynamics plays a central role in blade design. Engineers meticulously analyze airflow patterns around the blades using CFD simulations to minimize drag and maximize lift. This involves studying the wake – the turbulent air left behind by the blades – and its influence on subsequent blade passes. The wake can cause energy losses, so minimizing its effects is vital for optimizing turbine performance. Furthermore, the interaction between multiple blades and their wakes in a multi-blade rotor requires careful consideration to prevent negative interference effects. The goal is always to achieve a harmonious interaction between the blades and the surrounding air, resulting in smooth, efficient power generation.

One of the challenges in blade design lies in balancing performance with cost. While advanced materials and sophisticated designs can improve energy capture, they also increase manufacturing costs. Therefore, engineers must find an optimal balance between these factors, considering the trade-off between improved efficiency and increased initial investment. This optimization process involves comparing different designs and materials, analyzing their respective cost-benefit ratios, and selecting the option that provides the best return on investment over the turbine's operational lifetime. Sophisticated economic modeling is crucial in making informed decisions regarding blade design and material selection. Ultimately, the design process aims to maximize the energy output over the lifespan of the turbine while remaining cost-effective. This involves considering factors such as maintenance requirements, potential for repairs, and expected longevity of the components.

3.2 Gearboxes and Generators

The gearbox in a wind turbine is a crucial component, responsible for increasing the rotational speed of the low-speed shaft coming from the rotor to the speed required by the generator. This speed increase is necessary because wind turbines generally operate at low speeds, while generators are most efficient at higher speeds. Imagine trying to use a bicycle to power a washing machine; you'd need some serious gearing to make it work. The gearbox accomplishes this vital step in the energy conversion process. Different gearbox types exist, including planetary gearboxes, which are known for their high power density and reliability, and parallel shaft gearboxes, which are often chosen for their simplicity and lower cost. The selection of the right gearbox depends on several factors, including the size of the turbine, the desired generator speed, and the overall cost-effectiveness of the system. Careful consideration must be given to factors like efficiency, noise levels, and maintenance requirements.

Generators are the heart of the electricity production in a wind turbine; they convert the mechanical rotation from the gearbox into electrical energy. This is done through electromagnetic induction, a fundamental principle of physics. A rotating magnetic field within the generator interacts with stationary coils of wire, inducing an electric current. The voltage and frequency of this current depend on the speed of rotation and the generator's design. Common generator types used in wind turbines include induction generators and synchronous generators, each having its own advantages and disadvantages. Induction generators are generally more robust and simpler, while synchronous generators offer better control over the power output and frequency. The size and type of generator selected must carefully match the turbine's capacity and the requirements of the electrical grid.

The interplay between the gearbox and generator is a finely tuned dance of mechanical and electrical engineering. Poor design or inadequate maintenance in either component can significantly impact the overall efficiency and lifespan of the wind turbine. For instance, a malfunctioning gearbox can lead to increased wear and tear on other parts, or even catastrophic failure. Similarly, a poorly matched generator might not be able to effectively handle the power produced by the turbine, resulting in reduced energy output or instability in the grid. Therefore, ongoing monitoring and regular maintenance of both components are essential to ensure optimal performance and reliability of the entire wind energy system. Careful selection of components, considering their compatibility and robustness, is critical for maximizing the system's lifetime and return on investment.

Manufacturers are continuously improving the designs of gearboxes and generators, striving for greater efficiency, durability, and lower maintenance needs. Innovations include the use of advanced materials, such as high-strength steels and composites, to reduce weight and increase the component life. Furthermore, advancements in lubrication systems help minimize friction and wear, leading to increased efficiency and reliability. The integration of sophisticated monitoring and diagnostic systems allows for early detection of potential problems, preventing costly downtime. Regular inspections, coupled with predictive maintenance strategies based on real-time data analysis, are key to mitigating risk and ensuring the long-term operation of the wind turbine. Investing in high-quality components and implementing preventative maintenance practices is essential for the long-term profitability of any wind farm.

Understanding the interaction between the gearbox and generator within the broader context of the entire wind turbine system is key to comprehending its operation and optimizing its performance. It's not just about the individual components; it's about how they work together, sharing loads and transferring power. Designers must consider the overall system dynamics, including factors like wind variations, blade pitch control, and grid conditions, to ensure smooth operation and high efficiency across all parameters. The performance of the gearbox and generator directly impacts the overall energy yield of the wind turbine and its contribution to the sustainable energy mix. The selection of these components, therefore, is a critical decision that necessitates a thorough understanding of wind turbine technology, energy efficiency, cost-effectiveness, and ongoing maintenance practices. Choosing components that are optimized for performance, efficiency, and longevity is essential to ensure a cost-effective and reliable energy solution.

3.3 Control Systems

Efficient and reliable control systems are crucial for maximizing energy capture and extending the operational lifespan of wind turbines. These sophisticated systems continuously monitor various parameters, adapting turbine operation to optimize performance across diverse environmental conditions. They're not simply on/off switches; they're intricate networks balancing energy production with the structural integrity of the machine.

The core of any wind turbine control system lies in its ability to accurately measure wind speed and direction. Anemometers, strategically placed on the nacelle, provide this vital data. This data then feeds into a sophisticated algorithm; this algorithm, often implemented using embedded systems, makes real-time adjustments to blade pitch angles and rotational speed. Precision is paramount, as even small variations can significantly affect energy output and component stress.

Advanced control systems incorporate predictive capabilities, analyzing weather forecasts to anticipate shifts in wind conditions. This allows for proactive adjustments, smoothing out fluctuations in power generation and preventing abrupt changes that could damage the turbine. This predictive modeling minimizes wear and tear on components, leading to longer maintenance intervals and reduced downtime. By anticipating potential issues, these systems contribute significantly to the overall economic viability of the wind farm.

Beyond wind speed and direction, these systems monitor a vast array of data points. These range from generator temperature and oil pressure to blade vibrations and structural loads. Real-time monitoring of these critical parameters allows for early detection of potential problems. This allows for preventative maintenance, preventing minor issues from escalating into major failures. This proactive approach greatly reduces costly repairs and unscheduled outages. Sophisticated diagnostic tools within the control system assist technicians in identifying and addressing problems swiftly.

Effective control systems must also manage the power generated by the turbine, ensuring its safe and efficient integration into the electrical grid. This necessitates complex power electronics, which regulate voltage and frequency to match grid specifications. These systems smoothly transition between different power levels, maintaining grid stability, and preventing potentially damaging surges. Moreover, they play a key role in maximizing energy capture by balancing power generation with available grid capacity.

Safety is a top priority in wind turbine control systems. Emergency shutdown mechanisms are integrated to automatically halt operation in hazardous conditions, such as extreme winds, ice build-up, or internal component malfunctions. These safeguards are designed to prevent accidents and protect both the equipment and any personnel nearby. Furthermore, these systems often incorporate features that facilitate remote diagnostics and troubleshooting, allowing technicians to identify and address issues before they escalate.

The design and implementation of wind turbine control systems are constantly evolving. The incorporation of advanced algorithms, artificial intelligence, and machine learning techniques promises even greater efficiency, reliability, and safety in the future. These developments are driving further improvements in energy capture, reduced maintenance needs, and enhanced integration into smart grids. Innovations continuously push the boundaries of what is possible, unlocking greater potential from this critical renewable energy source.

Furthermore, the increasing complexity of wind turbine control systems necessitates robust cybersecurity measures. Protection against cyberattacks is vital to ensure the continued reliable operation of these critical infrastructure components. The secure transmission and processing of data are paramount to maintaining both the safety and economic viability of wind farms. Regular software updates and rigorous security protocols are essential to mitigate these potential risks.

The economic benefits of sophisticated control systems extend beyond reduced maintenance costs. By optimizing energy capture and improving grid integration, these systems contribute significantly to the overall profitability of wind farms. This increased efficiency helps to lower the levelized cost of energy, making wind power even more competitive as a sustainable energy source. The investment in advanced control systems ultimately pays for itself through enhanced performance and reduced operational expenses. It's a crucial element of the overall sustainability effort.

Modern wind turbine control systems are not just components; they are the nervous system of the entire operation. They gather, process, and respond to a constant stream of data, adapting in real-time to ever-changing conditions. Their capabilities extend far beyond simple power regulation; they are the backbone of safety protocols, diagnostic tools, and long-term operational efficiency. Without them, the widespread adoption and success of wind power as a renewable energy source would be seriously hampered.

The continued development and improvement of these systems are vital to the future of clean energy.

3.4 Tower Structures

The tower, a seemingly simple structure, is critical to a wind turbine's operation and longevity. Its design must balance several competing demands: withstanding immense forces from wind loads, supporting the heavy nacelle and rotor, and facilitating efficient energy transfer to the ground. Consider the forces involved—a substantial weight concentrated high above the ground, subjected to constant, dynamic pressures from the wind, and significant moments produced by the rotating blades. The design must account for these loads to prevent catastrophic failure.

Steel is the most common material for wind turbine towers, its high strength-to-weight ratio a key advantage. However, different steel grades and construction methods exist, each with specific properties. Lattice towers, once prevalent, are now less common due to their higher maintenance requirements compared to tubular towers, which are often constructed in sections for easier transportation and assembly. These sections are bolted or welded together on-site, a process requiring precise engineering and careful execution to ensure structural integrity. The choice of design depends on factors like turbine size, wind conditions, and soil characteristics at the installation site.

A critical aspect of tower design is its foundation. The foundation must anchor the tower securely, resisting overturning moments caused by wind loads and preventing settling over time. The type of foundation depends on the soil conditions; rocky terrain may permit shallower foundations, while softer soils require deeper, more complex designs such as piles driven deep into the ground. Proper foundation design is crucial for minimizing the risk of tower instability, a critical factor in the safety and operational reliability of the wind turbine. Insufficient foundation design can lead to long-term instability or even immediate failure in extreme conditions.

Beyond the fundamental structural aspects, other factors influence tower design. Considerations of accessibility for maintenance and repair are paramount. The tower's internal structure may need to accommodate access platforms, ladders, or even elevators, depending on the height and design. Furthermore, the tower's exterior needs to withstand harsh weather conditions, including corrosion, ice accumulation, and extreme temperatures. Corrosion protection, typically through painting or specialized coatings, is essential for extending the tower's lifespan, preventing costly repairs and downtime. The choice of materials and surface treatment significantly impacts the tower's long-term performance and maintenance needs.

Finally, the economic considerations of tower design and construction are equally important. Balancing structural integrity and cost-effectiveness is essential for the overall project feasibility. Optimizing the tower's design for minimal material usage without compromising strength can significantly reduce construction costs. Moreover, selecting readily available materials and construction techniques can further contribute to cost savings. The economic life cycle analysis, taking into account factors such as maintenance, repair, and eventual dismantling, is crucial in determining the overall efficiency and financial viability of the wind turbine project. A well-designed tower contributes to reducing long-term operational expenses.

3.5 Maintenance & Repair

Regular maintenance is paramount for maximizing the lifespan and efficiency of your wind turbine. Neglecting this crucial aspect can lead to costly repairs, reduced energy output, and even catastrophic failures. A proactive approach, incorporating scheduled inspections and preventative measures, is far more economical than reactive emergency fixes. This involves a detailed understanding of each component, its potential failure points, and the best practices for addressing them. We're talking about minimizing downtime and maximizing the return on your investment.

Preventive maintenance typically involves visual inspections of all major components. This includes examining the blades for cracks, erosion, or damage from lightning strikes, meticulously checking the gearbox for leaks or unusual noises, and thoroughly inspecting the generator for signs of overheating or wear. Beyond a visual inspection, more complex checks require specialized tools and expertise. These might include analyzing vibration patterns to detect early signs of bearing failure or using infrared cameras to identify hotspots that could indicate impending electrical problems. Remember, early detection is key to preventing major problems.

Cleaning the blades is also a significant aspect of preventative maintenance. Accumulated dirt, dust, and grime can reduce the aerodynamic efficiency of the blades, decreasing the amount of energy harvested. This reduction in efficiency translates to a direct loss of revenue. Regular cleaning, depending on the environment, might mean a simple washdown or a more thorough cleaning process using specialized equipment. The frequency of cleaning will depend heavily on environmental conditions – a coastal turbine will need more frequent cleaning than one located in a dry, desert climate.

Repairing damaged components requires specialized skills and equipment. Replacing a broken blade, for instance, is a complex operation involving specialized lifting equipment, precise alignment techniques, and a thorough understanding of the blade's aerodynamic properties. Repairing gearboxes often necessitates disassembly, inspection, replacement of worn or damaged parts, and meticulous reassembly. Generator repairs can be similarly involved, requiring testing and potentially replacement of individual components or even the entire unit.

The severity of any needed repair significantly impacts downtime. Minor repairs, such as tightening loose bolts or replacing small parts, might only take a few hours, causing minimal disruption to energy generation. Major repairs, such as blade replacement or gearbox overhaul, however, can take several days or even weeks. Minimizing downtime is crucial, necessitating efficient logistics, readily available spare parts, and a well-trained maintenance team. Planning for and anticipating potential problems is vital for managing unexpected occurrences.

Effective maintenance planning involves scheduling regular inspections and creating a detailed maintenance log. This log should document all inspections, repairs, and parts replacements, offering valuable data for predictive maintenance strategies. By analyzing historical data, patterns of component failures can be identified, allowing for proactive interventions and preventing future breakdowns. Regular preventative maintenance can significantly reduce the incidence of catastrophic failures, thereby protecting both the investment and the environment. This proactive approach not only prolongs turbine lifespan, but also minimizes safety risks associated with unexpected equipment failure.

Furthermore, a comprehensive training program for your maintenance team is vital.

This isn't just about basic mechanical skills. They need to understand the intricacies of wind turbine systems, including the complexities of hydraulics, pneumatics, and electrical systems. Safety training is also crucial; wind turbines are inherently hazardous pieces of equipment, and a lack of proper training can lead to serious accidents. Regular refresher courses keep everyone's knowledge up-to-date and ensure consistent, safe, and efficient maintenance practices. Investing in training is an investment in the long-term health and efficiency of your wind farm.

Finally, consider the environmental impact of your maintenance procedures. Proper disposal of used parts is crucial to minimize environmental contamination. Choosing environmentally friendly cleaning agents and lubricants minimizes the potential for environmental harm. Sustainable maintenance practices aren't just good for the environment, they also enhance your operation's reputation and can appeal to investors and the wider community. Responsible maintenance is an integral part of sustainable energy generation.

4 Chapter 4: Siting and Planning

Effective wind farm siting hinges on meticulous site assessment. This involves detailed analysis of wind resource characteristics—crucially, average wind speed, wind shear, and turbulence intensity. Sophisticated meteorological mast data, gathered over an extended period, are essential for accurate estimations. We need to identify areas with consistent, high wind speeds to maximize energy output, while simultaneously avoiding sites with excessive turbulence that can damage turbines or reduce their efficiency. Technological advancements in remote sensing and computational fluid dynamics now offer improved precision in modeling wind patterns, allowing for more precise site selection and optimized turbine placement. Careful consideration of terrain features is paramount; hills and ridges can significantly influence wind flow, creating areas of high wind speed but also potential obstructions. Analyzing detailed topographical maps and using advanced wind resource mapping software are crucial steps in effective site selection. Additionally, micro-siting within a selected area can further refine turbine placement, maximizing energy capture and minimizing negative impacts on neighboring communities. Such strategies lead to cost-effective and high-performing wind farms.
Environmental impact assessment is equally vital. Before any construction begins, a

comprehensive study needs to be conducted to assess potential ecological effects. This includes surveying the area to identify any sensitive habitats, such as bird migration routes or protected areas. Noise pollution, a significant concern for nearby residents, needs thorough investigation. Detailed acoustic modeling can predict noise levels and help design noise mitigation strategies, such as strategic turbine placement or noise barriers. Moreover, the visual impact on the landscape is a crucial consideration. Wind turbines, especially large ones, are inherently visible, leading to potential aesthetic concerns for some stakeholders. Proper visual impact assessments are needed, incorporating viewshed analysis and public participation to ensure community acceptance. These assessments inform the design and layout of the wind farm, possibly involving innovative strategies for visual integration with the surrounding landscape. Addressing these concerns early on contributes to obtaining necessary permits and securing community support for the project.

Careful consideration of regulatory frameworks is paramount for successful wind farm planning. The permitting process varies greatly depending on the location, and navigating these regulations efficiently is crucial. This includes understanding local zoning ordinances, environmental regulations, and bird and bat protection laws. Early engagement with relevant regulatory bodies, like local government agencies and environmental protection organizations, is vital for securing approvals and permits in a timely manner. Furthermore, obtaining community support is not just a recommendation; it's critical for the long-term success of any wind farm project. Effective community engagement involves proactively informing local residents about the project, addressing their concerns, and providing opportunities for feedback and participation. This can involve organizing public meetings, creating informational websites, and developing effective communication channels to keep the community well-informed and ensure their concerns are heard and addressed, leading to smoother project approvals and reduced potential conflicts.

Grid connection and energy transmission are critical factors in site planning. The proximity of the site to existing transmission lines significantly impacts the overall cost and feasibility of the project. Connecting to the grid involves working with electricity grid operators, which necessitates understanding grid capacity, voltage levels, and network stability. The complexity of grid connection procedures and costs vary depending on the scale of the wind farm and the existing grid infrastructure. Effective planning ensures seamless integration with the existing grid, minimizing potential disruptions and ensuring reliable energy delivery. Detailed grid studies, undertaken in coordination with the grid operator, help in determining the optimal grid connection points and the design of necessary substations and transmission lines. This stage is highly technical and requires experienced engineers who can navigate the complexities of power system analysis and grid integration. Proper planning minimizes costs and ensures stable power delivery to the consumers.

Finally, remember that financial aspects must be carefully planned. Wind farm development involves substantial upfront capital investment. Thorough cost estimations, including land acquisition, turbine procurement, construction, grid connection, and ongoing operation and maintenance, must be conducted. Developing realistic financial models and securing funding are essential. In this phase, exploring various funding options, such as government subsidies, private investment, or project financing, is paramount. Detailed economic viability assessments, considering factors such as energy production, electricity prices, and operating costs, are crucial to making informed decisions and ensuring long-term financial sustainability. This includes forecasting power output and revenue streams, taking into account factors such as wind resource variability and potential fluctuations in energy prices. A realistic and thorough financial plan greatly increases the probability of project success.

4.1 Site Assessment

Site assessment is paramount for successful wind farm development. It's not simply about finding a windy spot; it's a meticulous process requiring careful consideration of numerous factors, impacting the project's feasibility, efficiency, and environmental sustainability. Thorough investigation minimizes risks and maximizes returns, ensuring a financially viable and environmentally responsible energy project.

We begin with wind resource assessment. This involves measuring wind speed and direction at various heights above ground level for a prolonged period—at least a year is often recommended, to account for seasonal variations. Sophisticated meteorological instruments, like anemometers and wind vanes, are strategically positioned, typically on towers (meteorological masts) placed at potential turbine locations, providing crucial data on wind velocity, direction frequency, and turbulence intensity. This data is then analyzed using advanced statistical methods to generate a detailed wind resource map and to understand the wind patterns and power potential at the specific site. The quality and accuracy of this data significantly influence the subsequent design and profitability of the wind farm.

Next, we evaluate terrain characteristics. Topography greatly impacts wind flow, creating complex patterns that can enhance or diminish wind speed and introduce turbulence. Mountains, hills, valleys, and even forests can significantly alter wind profiles, thus necessitating accurate topographic mapping and wind flow modeling, ideally using advanced computational fluid dynamics (CFD) simulations to create detailed three-dimensional models of the airflow across the site, and to accurately predict wind resource availability at individual turbine locations. Understanding these complexities is critical for optimizing turbine placement and maximizing energy capture while minimizing stress on the turbines.

Environmental considerations are also key; they're integral to gaining social license and permitting. Conducting thorough environmental impact assessments (EIAs) is crucial. This involves identifying and evaluating potential impacts on local ecosystems and populations. These evaluations should include noise pollution modeling—using sophisticated software tools to predict the noise levels at nearby receptors. Studies on the potential visual impacts of the wind turbines on the landscape and potential effects on wildlife and habitats are also important aspects of the EIA. Understanding and mitigating these potential impacts early on will expedite the permitting process and contribute to a harmonious relationship with the local community and environment.

Infrastructure assessment is another critical element. Access to the site is paramount, including consideration of road networks, power transmission lines, and grid connection points. The site's proximity to existing infrastructure can significantly reduce construction costs and streamline the interconnection process, influencing both the project's budget and timeline. A thorough analysis of the existing infrastructure is needed. Moreover, the availability of skilled labor and local support services should also be considered. These factors can heavily influence the overall cost-effectiveness of construction and operations.

Finally, regulatory compliance is non-negotiable. Each location has specific regulations and permitting requirements. Navigating these procedures requires careful planning and consultation with relevant authorities. Thorough preparation, including environmental impact assessments and adherence to all local and national regulations, is paramount for obtaining the necessary permits and licenses, enabling project execution without legal setbacks. Proactive collaboration with regulatory agencies throughout the process can facilitate smooth approval and contribute to the project's timely completion. Ignoring these aspects can lead to delays, increased costs, and even project termination. A comprehensive site assessment, therefore, is an indispensable preliminary step in any wind farm development project. Careful attention to these detailed considerations sets the stage for a profitable and environmentally sound project.

4.2 Environmental Impact

The environmental impact of wind energy projects is a complex issue demanding careful consideration. While often touted as a clean energy source, wind farms aren't without consequences for the surrounding ecosystem. Careful planning and mitigation strategies are crucial to minimize these effects and ensure responsible development. Understanding these impacts is essential for the sustainable integration of wind energy into our energy portfolio.

Noise pollution is a primary concern. Turbine operation generates noise, impacting local wildlife and human residents. The intensity and range of this noise are dependent on several factors: turbine size, wind speed, distance from the turbine, and the surrounding topography. Studies have shown that certain bird and bat species are particularly sensitive to these noises, and careful acoustic modeling and noise barriers can help to reduce its effects. This includes choosing appropriate turbine locations and implementing effective noise reduction technologies during the planning stages. Understanding and mitigating this impact is crucial for community acceptance and the protection of wildlife.

Visual impacts are another significant consideration. The large scale of wind turbines can significantly alter the landscape, potentially impacting scenic views and affecting property values. This is especially true for areas of high aesthetic value, like those near national parks or along coastlines. Careful site selection, using turbines that blend visually with their surroundings, and considering the viewpoints of local communities are therefore crucial in reducing the negative impact. For example, some designs aim for camouflage using natural colours and patterns, and others focus on minimizing their visual footprint through optimized placement.

Wildlife interactions constitute another key environmental consideration. Collisions between birds and bats and the turbine blades are a well-documented concern, though the true extent of this impact remains under debate. These collisions are more likely during periods of high migration and low-light conditions, and the magnitude varies significantly depending on species and the location of the wind farm. Minimizing impacts requires carefully choosing turbine placement, considering migratory patterns, employing bird and bat deterrent technologies, and establishing monitoring programs. Data on collisions is essential to informing responsible siting practices and guiding the development of better mitigation strategies.

Furthermore, wind farm construction can cause habitat loss and fragmentation. The construction process involves land clearing, road building, and the installation of infrastructure, which can directly disrupt habitats and fragment ecological communities. Minimizing this impact necessitates thoughtful planning to protect sensitive ecosystems, using minimal land clearing wherever possible, restoring affected habitats, and using eco-friendly construction materials. Prioritizing these steps can significantly reduce the negative ecological consequences of wind farm construction.

The effects on water resources are another aspect to consider. Wind farms located near bodies of water can potentially impact water quality and aquatic life through runoff, erosion, and changes in water flow patterns during the construction phase. Moreover, the construction of access roads can lead to increased sedimentation and erosion, affecting aquatic ecosystems. Minimizing these effects requires careful consideration of water management plans, implementing erosion control measures, and selecting environmentally responsible construction techniques. The use of sustainable construction materials and practices are essential in achieving this goal.

Finally, the impact on soil and land use should be carefully addressed. The construction process can lead to soil compaction, erosion, and habitat disturbance. Effective soil management strategies are therefore essential, including minimizing ground disturbance, using appropriate soil stabilization techniques, and selecting suitable areas for construction to reduce the impact on land use. Careful restoration practices are crucial in minimizing the long-term consequences, while integrated land management planning for wind farm sites ensures the long-term health of the environment. Responsible development must carefully balance energy needs with ecological considerations to create a truly sustainable and environmentally sound energy infrastructure.

5 Chapter 5: Grid Integration

Connecting a wind farm to the power grid isn't simply plugging it in; it's a complex undertaking demanding careful planning and sophisticated technology. Initial steps involve extensive grid studies, analyzing the existing infrastructure's capacity and identifying the optimal connection point. This assessment considers factors such as transmission line capacity, voltage levels, and the overall stability of the grid. A detailed feasibility study is crucial, evaluating the technical and economic viability of the connection project. This includes assessing potential upgrades or expansions to the grid infrastructure needed to accommodate the incoming wind power.

The intermittency of wind power poses a significant challenge. Unlike conventional power plants, wind turbines don't generate a constant stream of electricity; their output fluctuates with wind speed. This variability can impact grid stability, potentially leading to voltage fluctuations or frequency deviations. Therefore, robust grid management strategies are essential to mitigate these effects. These strategies often involve sophisticated forecasting models that predict wind power output, allowing grid operators to anticipate and adjust power generation from other sources accordingly.

Smart grid technologies play a vital role in seamlessly integrating wind power. These technologies use advanced sensors, communication networks, and data analytics to optimize grid operations. Advanced metering infrastructure (AMI) provides real-time data on energy consumption and generation, allowing for efficient load balancing and dynamic dispatch of power. Furthermore, demand-side management (DSM) programs can incentivize consumers to shift energy usage to periods of higher wind generation, further improving grid stability. These integrated systems allow for efficient energy distribution and minimize potential disruptions.

Power output from wind farms needs meticulous management. To ensure smooth integration, wind turbines often incorporate sophisticated control systems capable of adjusting their power output in response to grid signals. These systems ensure that the turbines contribute stably to the grid, maintaining voltage and frequency within acceptable limits. Moreover, grid operators utilize various control strategies, such as frequency regulation and voltage support services, to maintain grid equilibrium even under fluctuating wind conditions. The goal is to make wind power a reliable and predictable source of electricity.

Beyond the technological aspects, effective grid integration also requires collaboration and coordination between various stakeholders. This includes wind farm developers, grid operators, regulators, and policymakers. Clear regulations and standards are necessary to ensure the safe and efficient integration of wind power. Open communication and data sharing between stakeholders are also crucial for optimal grid management and forecasting. Collaboration encourages innovation and addresses any emerging challenges that may arise during the integration process. Without careful coordination, integrating intermittent power sources can destabilize the grid, leading to brownouts or blackouts.

The specific connection methods vary depending on the scale of the wind farm and the existing grid infrastructure. Smaller wind farms might connect directly to lower-voltage distribution lines. Large-scale wind farms, however, often require dedicated transmission lines connecting them to high-voltage substations. Choosing the most appropriate connection method requires careful consideration of the technical, economic, and environmental factors. For example, using existing transmission lines may be more cost-effective but might necessitate upgrades. Building new transmission lines could be more expensive but would potentially offer greater flexibility and capacity.

Beyond direct connection, other integration strategies, such as using energy storage systems, are gaining increasing importance. Batteries or pumped hydro storage can store excess wind energy generated during periods of high wind speed and release it during periods of low wind or peak demand. This helps to smooth out power output fluctuations and further improves grid stability. However, implementing these storage systems entails further considerations regarding cost, scalability, and environmental impacts. Yet, their long-term potential is widely recognized as crucial to fully harnessing the advantages of renewable energy sources.

The ongoing evolution of grid technologies continues to enhance the integration of wind energy. Advancements in power electronics, smart sensors, and artificial intelligence are leading to more efficient and reliable grid management systems. These developments facilitate the integration of ever-increasing quantities of renewable energy, promoting a cleaner and more sustainable energy future. By continually adapting and improving grid infrastructure and control strategies, we can ensure the smooth and efficient integration of wind energy, paving the way for a greener and more resilient energy supply. This dynamic field demands continuous research and development, ultimately benefiting both the environment and the reliability of the global power grid.

5.1 Connecting to the Grid

Connecting wind turbines to the electrical grid is a complex process requiring careful planning and execution. The process begins with a comprehensive feasibility study, assessing the wind farm's capacity and the grid's ability to accommodate the influx of power. This involves detailed analysis of the existing grid infrastructure, including transmission lines, substations, and the overall grid capacity. Understanding the grid's limitations is crucial for preventing potential instability or overloading.

The next step involves the design and construction of the connection infrastructure. This typically includes new transmission lines or upgrades to existing ones, capable of carrying the substantial amount of power generated by the wind farm. Substations, essential for stepping up the voltage of the generated electricity for efficient long-distance transmission, might also need to be built or upgraded. Consideration must be given to the location and capacity of these substations, optimizing for minimum losses and maximum efficiency.

During the construction phase, rigorous safety protocols must be maintained. Coordination with utility companies is critical to ensure seamless integration into the existing grid. This collaborative effort requires effective communication, precise scheduling, and adherence to stringent safety standards. Detailed engineering plans, regularly reviewed and updated, are crucial for successful construction and grid integration. Moreover, the physical connection to the grid is a delicate operation, requiring specialized equipment and highly skilled technicians.

Once the physical connection is complete, thorough testing and commissioning are vital. This phase involves comprehensive testing of the entire system, ensuring all components are functioning as designed and that the wind farm can safely and reliably deliver power to the grid. Stringent quality control measures are implemented throughout this stage, addressing any issues promptly before full operational status. It is during this period that the system undergoes rigorous performance checks, identifying any potential issues related to power output or stability.

The ongoing monitoring and maintenance of the grid connection are equally important. Sophisticated monitoring systems track power output, voltage levels, and other critical parameters, providing early warnings of potential problems. Regular inspections and maintenance activities ensure the long-term reliability and safety of the connection. A proactive approach to maintenance minimizes downtime and prevents unexpected disruptions to power flow.

Finally, successful grid integration often requires navigating regulatory hurdles. Permits and approvals are needed from various regulatory bodies to ensure compliance with safety and environmental regulations. This involves interaction with governmental agencies and careful documentation of the entire process. Maintaining compliance with evolving standards and regulations is crucial for the long-term success of the wind farm project. These regulatory processes can be time-consuming but are essential for ensuring the safety and environmental soundness of the project. It's a testament to the careful consideration that goes into these projects.

The economic implications of grid connection should not be overlooked. The cost of building the necessary infrastructure can be substantial and needs to be carefully factored into the project's overall budget. Securing financing for these projects often requires detailed financial modelling and demonstrating the long-term economic viability of the wind farm. Furthermore, effective management of these costs is vital for successful project delivery and long-term profitability.

Beyond the financial aspects, integrating wind power into the grid contributes to broader environmental goals. By reducing dependence on fossil fuels, wind energy helps mitigate climate change and improves air quality. The successful integration of wind power into the grid signifies a significant step towards achieving a more sustainable and environmentally friendly energy future. This transition requires careful planning, investment, and a commitment to responsible energy generation.

The integration process also impacts the overall efficiency and stability of the power grid. Fluctuations in wind speed can affect power output, requiring sophisticated control systems to manage these variations and ensure grid stability. This demands the use of advanced grid management technologies, capable of handling intermittent power sources. Investing in robust control systems is crucial for maintaining grid stability and reducing the risks of outages. The efficiency of this process is crucial to the financial viability of the energy output.

In conclusion, connecting wind farms to the grid is a multi-faceted undertaking that demands careful planning, skillful execution, and ongoing monitoring. From initial feasibility studies to ongoing maintenance, each stage requires expertise and attention to detail. The successful integration of wind energy into the grid is essential for a sustainable energy future, offering a clean and renewable source of electricity while challenging the energy sector to innovate and adapt. This successful connection requires a holistic approach, considering technological, economic, and environmental factors. Ultimately, this process underscores the significance of wind power in the broader transition towards sustainable energy systems.

5.2 Power Output & Stability

Consistent power output is crucial for wind turbines to effectively contribute to the electricity grid. Fluctuations in wind speed directly impact the amount of electricity generated, creating variability that grid operators must manage. Understanding and mitigating these fluctuations is critical for reliable power delivery. This necessitates sophisticated control systems and, in many cases, energy storage solutions to buffer the intermittent nature of wind energy.

Predicting wind speed and direction with accuracy is a continuous challenge. While advanced meteorological models and forecasting techniques are improving, inherent uncertainties remain. These uncertainties necessitate the design of wind turbines and wind farms that can adapt to changing wind conditions. Furthermore, the design needs to incorporate redundancy and fail-safes to prevent cascading failures and ensure the continued operation of other turbines in the farm even when one or more fail.

Turbine placement plays a significant role in optimizing power output. Careful site selection, as detailed in Chapter 4, is paramount. Factors to consider include wind resource availability, proximity to transmission lines, and environmental impacts. Positioning turbines strategically within a wind farm to minimize wake effects—the disruption of airflow behind turbines—is key to maximizing energy capture across the entire farm. Sophisticated modeling and simulation tools help optimize turbine layouts.

Power output is not solely determined by wind speed; it's also significantly affected by turbine design and technology. As explored in Chapter 3, blade design, generator efficiency, and gearbox technology influence the overall energy conversion process. Advancements in blade aerodynamics continue to improve energy capture from even moderate wind speeds. Meanwhile, more efficient generators and gearboxes reduce energy loss during the conversion process.

Grid integration significantly influences the stability of power delivery from wind farms. Direct connection to the grid can introduce voltage fluctuations and frequency variations. Power electronic converters, often incorporated into wind turbines, play a vital role in smoothing out these fluctuations and maintaining grid stability. These converters help regulate the flow of electricity, ensuring smooth integration of the intermittent wind power into the overall grid operation.

Smart grid technologies, as discussed in Chapter 5, offer advanced solutions for managing wind energy's intermittency. These technologies facilitate better forecasting of wind power output and provide tools for balancing supply and demand across the grid. Real-time monitoring of wind farm performance, combined with predictive algorithms, allows for proactive management of power flow, reducing the impact of unpredictable wind patterns.

The economics of wind energy are closely linked to power output and stability. Consistent energy generation translates to greater revenue streams for wind farm operators. Conversely, periods of low power output or unexpected outages can severely impact financial returns. Therefore, maximizing power output and ensuring grid stability is essential for the overall financial viability of wind energy projects. This also affects the feasibility of future wind energy investment.

Furthermore, power output stability is crucial for meeting renewable energy mandates and carbon reduction goals. Integrating wind energy into the grid in a reliable and stable manner directly supports policy objectives aimed at decreasing greenhouse gas emissions. By addressing these stability issues, we pave the way for higher renewable energy penetration and a faster transition to sustainable energy systems.

Finally, ongoing research and development focus on improving power output and stability in wind energy systems. This includes innovations in turbine design, control algorithms, and grid integration strategies. Advancements in materials science and manufacturing processes are also leading to more durable and efficient wind turbines. The continuous development of hybrid power systems, often combining wind power with other renewable sources like solar, further enhances the reliability and predictability of energy supply. These advances strengthen the position of wind energy as a reliable component of future energy systems.

5.3 Smart Grid Technologies

Smart grids are crucial for effectively integrating wind energy into the power system. Their ability to manage the intermittent nature of wind power is paramount to ensuring grid stability and reliability. Without smart grid technologies, the fluctuations in wind power output could lead to significant challenges, potentially causing instability and blackouts.

Unlike traditional grids, smart grids employ advanced sensors, communication networks, and control systems to monitor and manage electricity flow in real-time. This real-time monitoring allows for immediate adjustments to the grid's operation in response to changes in wind power generation. This dynamic capability is unlike older systems that reacted more slowly to changes in supply and demand.

One key aspect of smart grid integration is advanced forecasting. Sophisticated weather models and data analysis predict future wind power output. These predictions help grid operators anticipate fluctuations and adjust power generation from other sources accordingly. This proactive approach minimizes disruptions and improves grid stability. Accurate forecasting, however, requires sophisticated algorithms and reliable weather data.

Another critical component is demand-side management. Smart grids can communicate directly with consumers, encouraging them to shift their energy consumption to times when wind power is abundant. This shift in demand helps balance supply and demand, reducing reliance on fossil fuel backups. For example, smart appliances can be programmed to run during periods of high wind energy production.

Energy storage systems play a vital role in stabilizing wind power's intermittent nature. Smart grids facilitate the integration of various storage technologies such as batteries and pumped hydro, allowing them to store excess energy during periods of high wind and release it when wind generation is low. This storage capability smooths out fluctuations and ensures a consistent power supply.

The communication infrastructure within a smart grid is essential. Advanced communication networks facilitate real-time data exchange between wind farms, grid operators, and consumers. This constant flow of information allows for efficient management of the entire system. Reliable and secure communication is therefore critical for effective grid management and preventing cyberattacks.

Smart grid technologies also enable better grid management by optimizing electricity distribution. They intelligently route electricity based on real-time conditions, minimizing transmission losses and ensuring efficient power delivery. This optimized routing prevents congestion and improves the overall efficiency of the power system. These strategies involve complex algorithms that continuously analyze grid conditions.

Furthermore, smart grids enhance the reliability and resilience of the power system.

They can rapidly detect and isolate faults, minimizing the impact of outages. This ability to identify and quickly react to faults enhances grid stability and reduces downtime, something crucial for a continuous energy supply. Advanced monitoring and self-healing capabilities are core elements of this resilience.

However, implementing smart grid technologies requires significant investment. Upgrading existing infrastructure and integrating new technologies can be costly. Furthermore, ensuring the security of these complex systems is critical to prevent cyberattacks and data breaches. Therefore, careful planning and investment strategies are needed to balance cost and effectiveness.

Finally, the successful integration of smart grid technologies depends on collaboration between stakeholders, including utilities, grid operators, and policymakers. A coordinated effort is essential to establish standards, regulations, and incentives that encourage the widespread adoption of smart grid technologies. Open communication and shared data are keys to coordinating the various actors within the system. This requires policy changes and international collaboration in some cases.

6 Chapter 6: Economic and Policy Aspects

The economic feasibility of wind energy projects hinges on several key factors. Initial capital costs, encompassing turbine purchase, installation, and land acquisition, represent a substantial upfront investment. Ongoing operational expenses, including maintenance, repairs, and insurance, must also be factored into the overall cost equation. These costs vary significantly depending on turbine size, technology, and location, necessitating a thorough cost-benefit analysis tailored to each specific project. This analysis should incorporate projected energy output, electricity prices, and potential revenue streams over the operational lifespan of the wind farm, typically twenty to thirty years. Furthermore, the analysis needs to account for potential revenue from carbon credits or other environmental benefits. A sophisticated financial model, capable of handling uncertainties and incorporating various scenarios, becomes essential for accurate assessment.

Government policies play a pivotal role in shaping the economic landscape of the wind energy sector. Many countries implement incentive programs designed to stimulate wind power development. These incentives can include direct subsidies, tax credits, and accelerated depreciation allowances, making wind energy projects more financially attractive. Renewable portfolio standards (RPS), which mandate a certain percentage of electricity generation from renewable sources, create a reliable market demand for wind energy, offering price stability and predictability. However, the specific types and levels of incentives can vary considerably between countries and regions, influencing the investment decisions of developers and impacting project profitability. Understanding these policies is crucial for accurately projecting the financial viability of wind energy projects in different geographical locations.

Market trends in the wind energy sector reflect a dynamic interplay of technological advancements, policy changes, and fluctuating energy prices. The global wind energy market has experienced considerable growth over the past decades, driven by increasing concerns about climate change and the need for cleaner energy sources. This growth has led to economies of scale, resulting in a gradual decline in the cost of wind turbines and other related technologies. Furthermore, technological advancements continuously improve turbine efficiency and energy capture capabilities, further enhancing the economic competitiveness of wind energy. However, market fluctuations, influenced by factors such as energy prices, government policies, and technological advancements, need to be carefully considered when assessing the long-term economic prospects of wind energy investments. Understanding the trends and forecasting future developments can provide investors with valuable insights into the likely profitability and returns on their investments.

Energy policies, both at national and international levels, significantly influence the deployment and growth of wind energy. Policies addressing climate change often set ambitious targets for renewable energy penetration, driving investment in wind power. National energy strategies may include specific targets for wind energy capacity or renewable energy mixes. International agreements and collaborations can also play a crucial role in promoting the global adoption of wind energy. However, inconsistent or fluctuating government policies can create uncertainty and risk for investors. A stable and supportive policy environment is essential for attracting investment and ensuring the long-term growth of the wind energy sector. Furthermore, effective policy mechanisms must address grid integration challenges, ensuring the seamless and reliable integration of wind power into existing electricity grids.

The future of wind energy looks bright, with promising prospects for continued growth and technological innovation. Advances in turbine technology are continuously improving efficiency and reducing costs. The development of offshore wind farms offers substantial potential for large-scale energy generation in areas with consistently high wind speeds. Emerging technologies, such as floating offshore wind turbines and hybrid energy systems integrating wind and other renewable sources, are poised to further enhance the capabilities and competitiveness of wind energy. However, challenges remain, including the need for improved grid infrastructure and the mitigation of environmental impacts. Addressing these challenges through strategic policy initiatives and technological advancements will be vital for realizing the full potential of wind energy in meeting global energy demands and environmental goals. A future where wind power plays a central role in the global energy mix is increasingly likely as technologies continue to improve and policy support strengthens.

6.1 Cost Analysis

Understanding the financial aspects of wind energy projects is crucial for both developers and investors. A thorough cost analysis considers various factors, from initial investment to ongoing operational expenses, influencing the project's overall economic viability. Ignoring these details can lead to inaccurate estimations, jeopardizing the project's success.

Initial capital expenditures (CAPEX) represent the upfront costs associated with project development. These substantial sums cover land acquisition, permitting fees, turbine procurement, construction, and grid connection infrastructure. Accurate estimations require careful planning, detailed engineering designs, and competitive bidding processes for various components. Underestimating CAPEX can result in significant budget overruns, potentially derailing the entire venture. Therefore, comprehensive feasibility studies, including contingency planning for unforeseen costs, are essential. Precise estimations are critical for securing project funding and ensuring the project's financial sustainability.

Operational expenditures (OPEX) are the ongoing costs associated with maintaining and operating the wind farm throughout its lifespan. These expenses include routine maintenance (like blade cleaning and gearbox lubrication), repair costs for damaged components, insurance premiums, and administrative overheads. OPEX varies based on factors such as turbine technology, environmental conditions, and the availability of skilled maintenance personnel. Regular inspections and preventative maintenance strategies play a vital role in minimizing unexpected repair expenses, ensuring long-term operational efficiency and reduced OPEX. Detailed OPEX budgeting, considering potential inflation and spare parts costs, is crucial for long-term project profitability.

Energy production estimates are integral to the economic analysis. They depend on several factors, including the wind resource at the selected site, the turbine's capacity factor, and the overall efficiency of the wind farm's operation. Accurate predictions require sophisticated meteorological data analysis and advanced simulation modeling. The predicted energy production directly impacts the project's revenue projections and the evaluation of its financial feasibility. Underestimating energy production can lead to a significant drop in projected revenue and an inaccurate assessment of the project's return on investment. Therefore, accurate estimations, taking into account variations in wind speed and turbine performance, are crucial.

The financial models used to assess project viability incorporate all aspects of cost and revenue. These models account for CAPEX and OPEX, energy production estimates, and the predicted price of electricity over the wind farm's lifetime. The model often includes sensitivity analysis, examining the impact of variations in crucial parameters (such as electricity price and wind resource) on the project's profitability. A thorough financial model allows developers to evaluate the project's risk profile, assess its resilience to external factors, and make informed investment decisions. The choice of financial model should align with the specific circumstances of the project and its risk tolerance level. Transparency in the financial model is key to attracting investors and ensuring successful funding.

Government policies and incentives significantly influence wind energy project economics. Subsidies, tax credits, and renewable energy mandates can significantly reduce the initial investment costs and improve project returns. The availability of these incentives varies across regions and over time. Developers must carefully consider applicable regulations and available funding opportunities to optimize project economics. Thorough knowledge of government policies and regulations, including future policy changes, is critical for project feasibility analysis and long-term investment strategy. Furthermore, government support can significantly reduce the risks associated with renewable energy projects.

In conclusion, a comprehensive cost analysis is crucial for the financial success of any wind energy project. It requires careful consideration of various factors from initial investment costs to long-term operational expenses and government incentives. Accurate estimations, robust financial modeling, and a thorough understanding of relevant regulations are all essential elements of a successful cost analysis. A well-executed analysis minimizes risk, attracts investors, and ensures the long-term viability of wind energy projects as a sustainable solution for electricity generation. Failing to conduct a proper cost analysis can result in unforeseen expenses and potentially the failure of the project, ultimately hindering progress towards cleaner energy sources.

6.2 Government Incentives

Government incentives play a crucial role in shaping the economic landscape of wind energy projects. These incentives, often designed at national or regional levels, aim to stimulate investment, reduce financial risk for developers, and accelerate the adoption of wind power as a viable renewable energy source. Understanding these mechanisms is therefore essential for anyone involved in, or studying, the wind energy sector.

Subsidies directly reduce the initial cost of wind energy projects. These can take many forms, including direct grants, tax credits, or rebates offered per megawatt of installed capacity. For example, many countries offer production tax credits (PTCs) that provide a certain amount of money for each kilowatt-hour of electricity generated over a set period. This significantly reduces the financial burden on developers, making projects more attractive to investors. The impact is substantial: PTCs can significantly lower the levelized cost of energy (LCOE), making wind energy more competitive with fossil fuel alternatives.

Renewable portfolio standards (RPS) represent another influential category of government incentives. These mandates require electricity suppliers to generate a specific percentage of their power from renewable sources, such as wind, within a given timeframe. RPS mandates create a guaranteed market for wind energy, ensuring a steady stream of revenue for project developers. This predictability reduces risk and fosters long-term investment. The success of RPS programs, however, hinges on the stringency of the regulations and the enforcement mechanisms in place. A poorly designed or inadequately enforced RPS may fall short of its intended objective.

Tax incentives, encompassing a range of measures, further support wind energy development. Accelerated depreciation allows developers to write off a larger portion of the investment cost in the early years of operation. Investment tax credits provide a direct reduction in taxable income, similar to PTCs but often applied to the initial investment rather than ongoing production. These measures, while varying in their specifics across jurisdictions, all share the common aim of making wind energy projects more financially appealing. The complexity of tax codes, however, sometimes requires specialized financial expertise to effectively utilize these incentives.

Beyond direct financial support, governments can influence wind energy development through indirect measures. Streamlined permitting processes, for instance, reduce the time and cost associated with obtaining the necessary approvals for project construction and operation. This expedited approval process can significantly shorten the project timeline and lessen the associated financial risks. Likewise, investment in grid infrastructure upgrades, specifically to accommodate the intermittent nature of wind power, helps facilitate greater integration of wind energy into the existing power grid.

Additionally, governments play a key role in fostering technological advancements in wind energy. Funding research and development programs accelerates innovation, leading to the development of more efficient, durable, and cost-effective wind turbines and related technologies. Support for education and training programs creates a skilled workforce capable of designing, constructing, and maintaining wind energy systems. This ensures that the industry possesses the expertise necessary to realize its full potential. Government-supported educational initiatives equip the next generation with the necessary skills to innovate and sustain the long-term growth of the sector.

It's important to note that the effectiveness of government incentives is not uniform. Factors such as the specific design of the incentive, the economic and political context, and the overall regulatory environment all play significant roles in determining the success of these programs. Careful consideration of these contextual factors is essential in designing effective policies. The ultimate impact of any incentive scheme depends on its implementation and coordination with other policies. Robust monitoring and evaluation are critical for assessing the effectiveness of government interventions. Moreover, it's imperative to consider the potential unintended consequences, such as market distortions or inequities, and ensure that incentive programs are designed to promote fair competition and sustainable development. Therefore, a holistic and well-planned approach to government incentives is crucial for fostering a thriving wind energy industry.

6.3 Market Trends

Understanding the current market trends in wind energy is crucial for anyone involved in the industry, from investors to engineers. The sector's growth isn't uniform; certain segments experience explosive expansion while others face challenges. This dynamic landscape requires careful analysis and a forward-looking perspective.

Currently, offshore wind projects are witnessing a surge in investment and development. This is largely driven by the significantly higher wind speeds found further out at sea, leading to increased energy generation potential compared to onshore projects. Technological advancements, such as larger turbine designs and improved installation techniques, have also made offshore wind more economically viable. However, the high upfront capital costs and the complex logistics involved in offshore construction remain significant hurdles.

Onshore wind power, while more mature, continues to evolve. There's a growing trend towards larger wind turbine installations, particularly in areas with optimal wind conditions. This size increase translates into greater energy output per turbine, improving the overall efficiency and reducing the cost per unit of energy. However, this expansion isn't without its drawbacks; larger turbines require more extensive infrastructure and potentially lead to greater visual impact on the surrounding landscape.

Another notable trend is the increasing integration of wind energy with other renewable sources, creating hybrid energy systems. This diversification enhances grid stability and reduces reliance on any single source. Combining wind power with solar, for example, effectively balances the intermittent nature of both technologies, providing a more consistent and reliable power supply. These hybrid systems are particularly advantageous in locations with varying wind and solar resource availability.

Policy and regulatory landscapes significantly influence market trends. Government incentives and supportive policies, such as tax credits and renewable portfolio standards, have played a vital role in the growth of the wind energy sector. These incentives vary across regions and countries, creating a diverse market dynamic where some areas experience rapid growth while others lag behind. Understanding these regulatory differences is therefore critical for investors and developers.

The energy storage market is tightly coupled with wind energy's future. Wind power's intermittent nature necessitates effective energy storage solutions to manage fluctuations in output and ensure a stable power supply. Significant advancements in battery technology and other energy storage solutions are helping to address this challenge, improving the viability and reliability of wind power as a primary energy source. This interdependency is a key market factor to observe.

Furthermore, technological innovation is continually shaping the market. Improvements in turbine blade design, generator efficiency, and control systems are all contributing to better energy capture and lower operational costs. Research and development efforts focused on advanced materials, smart grid integration, and predictive maintenance techniques are also transforming the sector. Keeping abreast of these technological advancements is crucial for staying competitive.

Market competition is another crucial factor to consider. The wind energy sector is increasingly competitive, with established manufacturers and new entrants vying for market share. This competition drives innovation and cost reduction, benefiting consumers and facilitating wider adoption of wind power. Understanding the competitive landscape and the strategies of key players is essential for successful participation in this dynamic market.

Finally, public perception and acceptance are also factors influencing market trends. Growing public awareness of climate change and the need for renewable energy sources is driving increased support for wind power. However, concerns about environmental impacts, such as visual effects and potential harm to wildlife, need careful consideration. Successfully navigating these public perceptions is vital for the long-term success of wind energy projects.

The future of the wind energy market will continue to be driven by a complex interplay of technological advancements, supportive policies, fluctuating energy prices, and evolving public opinion. Careful analysis of these interacting factors is critical for understanding current market trends and predicting the future trajectory of this vital sector within the broader context of sustainable power generation. Staying informed and adaptable in this dynamic environment is key to success.

6.4 Energy Policy

Effective energy policy is crucial for realizing the full potential of wind energy. It's not merely about supporting the technology; it's about creating a supportive ecosystem that encourages investment, innovation, and responsible development. This necessitates a multi-pronged approach that addresses economic incentives, environmental regulations, and grid integration strategies. Without such comprehensive policies, the transition to a sustainable energy future powered by wind will be hampered.

Successful implementation requires a clear understanding of the economic realities. Wind energy projects, while environmentally beneficial, involve significant upfront capital investment. Government incentives, such as tax credits, subsidies, and feed-in tariffs, play a vital role in making these projects financially viable, particularly during the initial stages of development when returns are uncertain. Furthermore, policies must also account for the fluctuating nature of wind power, ensuring that the energy grid remains stable and reliable despite this intermittency.

Environmental regulations, especially concerning noise pollution and wildlife impact, are paramount. Careful siting of wind farms, taking into account migratory patterns of birds and bats, coupled with the use of quieter turbine technologies and mitigation strategies, can effectively minimize environmental effects. Equally important is addressing potential visual impacts on landscapes, which may require careful planning and public consultation. Striking a balance between sustainable energy generation and environmental protection is pivotal.

Grid integration is another critical aspect. Wind energy needs effective transmission infrastructure to deliver electricity to where it's needed. Policies should encourage investment in upgrading existing grids and constructing new ones to accommodate the influx of renewable energy. This also involves implementing smart grid technologies, which can enhance grid stability and manage the intermittency inherent in wind power generation. The ability to store surplus energy and effectively manage fluctuating outputs through technological advancements is crucial.

Policy should extend beyond immediate financial incentives and focus on long-term sustainability. Supporting research and development of cutting-edge wind turbine technologies, such as advanced blade designs and improved energy storage solutions, is essential to drive down costs and increase efficiency. This could take many forms, including grants for research institutions, incentives for private sector investment in research and development and the establishment of industry-academia collaborations. Such initiatives are key to ensuring that wind energy remains competitive and attractive in the long term, leading to technological advancements and increased efficiency.

International cooperation plays a crucial role. Sharing best practices, standardized regulations, and collaborative research efforts across borders can foster the wider adoption of wind energy. International agreements on carbon emission reduction targets, coupled with incentives for cross-border wind energy projects, can create a global market that promotes innovation and scalability. Effective policies must not be confined within national borders but should strive for global collaboration to maximize the impact of wind energy as a sustainable alternative to fossil fuels.

Furthermore, fostering public awareness and engagement is equally important. Education and communication campaigns that highlight the economic and environmental benefits of wind energy can build public support and encourage community involvement. Transparent and participatory planning processes, which involve stakeholders from all walks of life, can enhance the acceptance and success of wind energy projects. Addressing potential concerns and misconceptions, and explaining the long-term benefits in terms of job creation and economic growth, is essential for acceptance among the community and policymakers.

A robust policy framework should embrace a holistic approach, encompassing economic incentives, environmental protection, grid integration, research and development, and public engagement. It's a complex undertaking that demands a long-term vision and coordinated efforts from governments, industries, and communities. The benefits, however, are immense, leading to a cleaner, more sustainable energy future, powered by the wind. The success of wind energy, and its contribution to a greener world, is directly tied to the efficacy and foresight of the policy decisions made today. We must act decisively to ensure that wind energy achieves its full potential as a cornerstone of our sustainable energy future.

6.5 Future of Wind Power

Offshore wind presents a particularly exciting frontier. The higher and more consistent wind speeds found at sea offer significantly greater energy yields compared to land-based turbines. Technological advancements in floating platforms and robust turbine designs are making deep-water installations increasingly feasible, unlocking vast untapped resources. Imagine massive wind farms stretching across ocean horizons, silently generating clean energy on a scale previously unimaginable. This expansion isn't just about harnessing more power; it's about strategic resource allocation, mitigating land-use conflicts, and creating new economic opportunities in coastal communities. The challenge lies in addressing the unique logistical and environmental considerations inherent in offshore operations, from installation complexities to the potential impact on marine ecosystems. However, the rewards – abundant clean energy and a more sustainable future – far outweigh the obstacles.

Further advancements in turbine technology will also play a crucial role. Lightweight yet incredibly strong materials, like advanced composites, are continuously being developed, leading to larger, more efficient blades capable of capturing even the slightest breezes. Sophisticated control systems are being refined to optimize energy capture in fluctuating wind conditions, increasing overall efficiency and reducing energy losses. The integration of smart sensors and data analytics allows for predictive maintenance, minimizing downtime and maximizing turbine lifespan. These technological leaps aren't just incremental improvements; they represent a paradigm shift toward a more intelligent and responsive wind energy infrastructure.

The economic landscape of wind power is rapidly evolving. While initial capital costs remain substantial, ongoing technological innovation is driving down the price of wind energy, making it increasingly competitive with traditional fossil fuel sources. Government policies, including tax incentives and renewable energy mandates, play a vital role in fostering market growth and ensuring the long-term viability of wind power projects. The increasing awareness of climate change and the urgent need for clean energy alternatives are propelling investments in wind energy, creating a robust and expanding market for manufacturers, developers, and related industries. This economic dynamism is not only stimulating technological innovation but is also contributing to the creation of high-skill jobs in manufacturing, engineering, and operations.

However, the future of wind power isn't without its challenges. Grid integration remains a critical issue, requiring significant upgrades to existing infrastructure to accommodate the intermittent nature of wind energy. Effective energy storage solutions are crucial to mitigating fluctuations in power output and ensuring grid stability. Furthermore, the environmental impacts of wind farms, both onshore and offshore, require careful consideration and mitigation strategies. A holistic approach is essential, balancing the need for clean energy with the protection of ecosystems and biodiversity. Addressing these challenges proactively is crucial to unlocking the full potential of wind power and ensuring its sustainable integration into our energy systems.

Looking ahead, hybrid energy systems, combining wind power with other renewable sources like solar or hydro, represent a promising avenue for greater energy security and reliability. These integrated systems offer a robust and diverse energy portfolio, minimizing dependence on any single energy source and enhancing resilience to fluctuations in weather patterns. Integrating energy storage technologies, such as battery systems or pumped hydro, further enhances the flexibility and reliability of these hybrid systems.

This synergistic approach offers a pathway toward a truly sustainable and decentralized energy future, maximizing the benefits of diverse renewable resources. The development and implementation of such intelligent energy grids will be critical in shaping the future of wind power and enabling the transition to a cleaner, more sustainable energy landscape.

7 Chapter 7: Case Studies and Future Trends

The Horns Rev wind farm off the coast of Denmark stands as a testament to the potential of offshore wind energy. Its success, characterized by consistent high energy output and minimal environmental disruption, serves as a model for future projects globally. This farm's sophisticated grid integration systems, designed to handle the intermittent nature of wind power, have also demonstrated effective management of power fluctuations, minimizing impact on grid stability. The lessons learned from Horns Rev, including meticulous site selection, robust infrastructure design, and proactive maintenance strategies, are invaluable to developers of similar projects worldwide. Its long-term performance data provides critical insight into the long-term viability and economic benefits of large-scale offshore wind energy installations. The farm's experience showcases the synergy between technological innovation and effective resource management, driving efficiency and sustainability.

Let's shift our focus to emerging technologies poised to revolutionize wind energy. Advanced blade designs, incorporating lighter yet stronger materials like carbon fiber composites, are increasing energy capture efficiency. These newer blade designs exhibit improved aerodynamic properties, leading to greater energy yield from the same wind speeds. Simultaneously, these materials reduce wear and tear, extending operational lifespan and lowering maintenance costs. This combination of enhanced efficiency and durability significantly impacts the overall economic viability of wind energy projects. Furthermore, research into innovative control systems, utilizing artificial intelligence and machine learning algorithms, promises to optimize turbine performance and maximize energy production in variable wind conditions. This technology promises significant advancement in turbine efficiency and overall power output.

Offshore wind energy is rapidly gaining traction as a significant contributor to sustainable power generation. The deeper waters offer consistently higher wind speeds compared to onshore locations, resulting in greater energy output. However, developing offshore wind farms presents unique challenges, requiring specialized engineering, construction, and maintenance techniques, particularly considering the harsh marine environment. Subsea cable technology is constantly evolving to cope with the challenges of transmitting power from offshore turbines to land-based grid systems. These advancements, combined with improvements in floating turbine platforms suitable for deep waters, are expanding the geographical reach of offshore wind energy, tapping into vast untapped wind resources.

Several notable case studies illustrate successful wind farm implementations. The Alta Wind Energy Center in California, one of the largest onshore wind farms in the US, exemplifies effective large-scale deployment. Its success is attributable to careful site selection, meticulous environmental impact assessments, and sound project management. The farm's long-term performance data has proven valuable in shaping future onshore wind energy projects. Similar success stories from across the globe highlight the diverse adaptability of wind energy across varying geographical landscapes and climatic conditions, providing valuable data for future installations. These successful projects have demonstrated the potential of wind energy to contribute significantly to the global energy transition, prompting further investments and research.

Looking ahead, hybrid energy solutions are gaining prominence. Integrating wind energy with other renewable sources, such as solar power or hydroelectricity, enhances grid reliability and mitigates the intermittency of wind power. This strategy provides a more consistent and predictable energy supply, optimizing resource utilization and minimizing reliance on fossil fuels. Furthermore, the integration of energy storage technologies, like advanced battery systems or pumped hydro storage, effectively addresses the variable nature of wind energy, smoothing out fluctuations and ensuring a continuous flow of power to the grid. This integrated approach promises a more resilient and sustainable energy system. The synergy between different renewable resources results in a more stable and reliable power supply, improving energy security and sustainability.

The future of wind energy hinges on continuous innovation and technological advancements. Researchers are exploring new materials for turbine blades, aiming for lighter weight, increased durability, and enhanced aerodynamic performance. Developments in blade design have yielded a significant increase in energy efficiency, maximizing the energy harvested from each turbine. Furthermore, the integration of smart grid technologies and advanced control systems promises to optimize energy distribution and ensure grid stability even with fluctuating wind power input. These technological advancements work together to enhance both efficiency and reliability within the wind energy sector. This ongoing pursuit of technological improvements is a key driver for the sector's continued growth and expansion.

7.1 Successful Wind Farms

The Horns Rev wind farm, located off the coast of Denmark, stands as a testament to successful large-scale wind energy projects. Its strategic placement capitalizes on consistent, high-velocity winds prevalent in the North Sea, resulting in exceptional energy production. This achievement highlights the importance of meticulous site selection, a crucial factor often overlooked in early planning stages. The farm's success isn't solely attributable to location; it's a culmination of advanced turbine technology, robust grid infrastructure, and proactive maintenance strategies. Thorough environmental impact assessments, conducted before construction, mitigated potential disruptions to marine ecosystems. Such rigorous planning is fundamental to achieving long-term sustainability.

Beyond optimal siting, the Horns Rev project exemplifies the importance of technological innovation. The turbines employed incorporate cutting-edge designs that maximize energy capture even in challenging weather conditions. Advanced blade aerodynamics and efficient generator systems contribute significantly to overall efficiency. Furthermore, sophisticated control systems constantly monitor and adjust turbine operation, optimizing performance and minimizing wear and tear. This ongoing optimization directly translates into increased energy output and reduced operational costs, showcasing a key element for long-term financial success.

Effective grid integration is another critical aspect of Horns Rev's success story. The project's seamless connection to the Danish power grid ensures consistent energy delivery. The integration process itself wasn't straightforward; it required significant investment in grid infrastructure upgrades to handle the intermittent nature of wind power.

This involved developing sophisticated forecasting models to predict wind energy output accurately, allowing for better grid management and reducing the risk of power fluctuations. Advanced grid technologies enable efficient balancing of wind energy with other sources, ensuring stable power supply.

Financial viability is a crucial element in the long-term success of any wind farm. Horns Rev offers valuable lessons in managing the economic aspects of wind power generation. Initial capital investment in the project was substantial, encompassing turbine procurement, installation, grid connection, and environmental impact mitigation. However, the consistent and substantial energy output has yielded a strong return on investment, offsetting initial costs and providing considerable long-term profitability. Government subsidies and tax incentives initially played a role, fostering early development.

The operational phase requires ongoing investment in maintenance and repair. A robust maintenance program, incorporating both scheduled and reactive maintenance, ensures optimal turbine performance and longevity. This reduces downtime and maximizes energy generation, contributing directly to overall financial success. Horns Rev's experience highlights the need for a structured maintenance plan, including regular inspections, component replacements, and skilled technicians, all essential to reducing operational costs. It's a lesson in cost efficiency and long-term viability.

Furthermore, Horns Rev serves as an excellent case study for the successful integration of a large-scale renewable energy project into a community. Community engagement during the planning and development phases was critical, ensuring local acceptance and minimizing negative impacts. The project also created numerous local jobs, boosting the regional economy. Careful consideration of local concerns and collaboration with local stakeholders are key in navigating potential challenges and fostering public support. This success emphasizes the significant socioeconomic benefits attainable from responsible wind farm development. Public acceptance and support are vital for any renewable energy project to succeed.

Successful wind farm projects like Horns Rev demonstrate that meticulous planning, advanced technology, and proactive management are essential for long-term success. These projects highlight the importance of thorough site assessments, optimal turbine technology, seamless grid integration, and a strong financial model. Beyond these technical and economic aspects, responsible community engagement and a commitment to environmental sustainability are also crucial elements for achieving widespread acceptance and lasting impact. The Danish experience is a model for future wind energy projects worldwide. Thorough planning, including addressing the challenges of intermittency and grid stability, is key to a long-term, stable supply of electricity. The success of such projects rests not just on technical prowess but also on a balanced approach considering economic, social, and environmental factors.

7.2 Offshore Wind Power

Offshore wind power represents a significant advancement in harnessing wind energy's potential. Unlike onshore wind farms, which are often limited by land availability and public perception, offshore sites offer vast expanses of open water with consistently higher wind speeds and less visual impact on land-based communities. This translates to increased energy generation capacity and greater overall efficiency. However, the transition to offshore wind presents unique challenges.

The construction and maintenance of offshore wind farms are considerably more complex and expensive than their onshore counterparts. Specialized vessels, heavy-duty equipment designed for marine environments, and robust anchoring systems are crucial for installation in often turbulent and unpredictable waters. Furthermore, harsh weather conditions, including strong winds, high waves, and sea ice (in some regions), necessitate the use of materials and techniques able to withstand considerable stress and prolonged exposure to the elements. This added complexity leads to higher upfront capital costs.

Logistical considerations also increase the challenge. The transportation of massive turbine components to offshore locations requires specialized transport vessels, careful planning and coordination to avoid disruptions caused by weather and sea traffic, and well-defined safety protocols for all personnel involved. Once operational, maintaining and repairing these turbines requires specialized teams and equipment capable of reaching and working on the turbines despite adverse weather conditions. Regular inspections and preventative maintenance are essential to ensure long-term operational effectiveness and cost-effectiveness.

Despite these challenges, offshore wind farms boast several advantages. Their higher wind speeds result in significantly increased energy production per turbine compared to their onshore counterparts. These higher capacities can offset the higher initial investment costs over the lifetime of the project. Moreover, the large-scale deployment of offshore wind farms has the potential to contribute significantly to national and international renewable energy goals and reduce carbon emissions on a scale that would be difficult to achieve using onshore wind farms alone. This plays a crucial role in mitigating climate change and ensuring energy security.

The environmental impact of offshore wind farms needs careful evaluation. While concerns about potential effects on marine life, including noise pollution, habitat disruption, and bird collisions, must be addressed through careful site selection, environmental impact assessments, and mitigation strategies, numerous studies suggest that the impacts of well-planned and properly operated offshore wind farms are significantly lower compared to other energy sources. Sustainable practices, including minimizing impacts on marine ecosystems and developing effective bird and bat deterrent systems, are fundamental to responsible offshore wind development.

Technological advancements continuously improve the efficiency and reduce the cost of offshore wind energy. Improvements in turbine design, including longer blades and advanced control systems, enhance energy capture. Innovations in foundation design and installation techniques lead to reduced costs. The integration of smart grid technologies is critical for managing the intermittent nature of renewable sources such as offshore wind and ensures grid stability. Research into floating wind turbine technology opens up possibilities for exploitation of deeper water sites with stronger and more consistent winds, further expanding the potential of offshore wind energy.

Governments worldwide increasingly recognize the potential of offshore wind energy and implement supportive policies. These include subsidies, tax incentives, and regulatory frameworks designed to promote investment and development. International collaborations play a significant role in sharing best practices, fostering technological advancements, and coordinating offshore wind farm development across borders. The long-term economic viability of offshore wind energy is becoming increasingly attractive, driven by declining turbine costs, increasing demand for renewable energy, and supportive policy measures.

In conclusion, the transition to offshore wind represents a crucial step towards a sustainable energy future. While challenges related to cost, logistics, and environmental impact need careful consideration and mitigation strategies, the advantages of higher energy output, reduced land-use impact, and large-scale contribution to decarbonization outweigh the challenges. Continued technological innovation and supportive policies will play a significant role in unlocking the full potential of offshore wind power and driving its expansion in the years to come. The future of sustainable energy relies heavily on harnessing the significant potential of offshore wind. This is not merely a matter of technological advancement; it is a necessary step in securing a cleaner and more sustainable global energy supply.

7.3 Emerging Technologies

Offshore wind farms represent a significant advancement, pushing the boundaries of what's possible in wind energy generation. Their capacity to harness stronger and more consistent winds, compared to their onshore counterparts, leads to substantially higher energy yields. The challenges, however, are considerable, demanding robust turbine designs capable of withstanding the harsh marine environment—extreme weather, saltwater corrosion, and the added complexities of installation and maintenance in deep waters. These factors significantly impact project costs, necessitating innovative solutions to improve efficiency and reduce expenses.

Floating wind turbine technology is a particularly promising area. Unlike fixed-bottom structures, these floating platforms allow for deployment in significantly deeper waters, unlocking vast untapped wind resources. This opens up possibilities for massive wind farms in regions previously considered inaccessible. The engineering challenges are significant, requiring advanced mooring systems and dynamic control mechanisms to ensure stability and optimal energy capture, especially during stormy conditions. Research into novel materials, such as lighter yet stronger composites, is crucial to reducing costs and improving the long-term viability of floating wind farms.

Advanced blade designs are continuously being refined. The pursuit of maximizing energy capture from even the slightest breeze requires meticulous aerodynamic modeling and material science. New blade materials, including composites reinforced with carbon fiber, are being explored to enhance strength, reduce weight, and improve durability. This minimizes maintenance requirements and extends the operational lifespan of turbines. Moreover, the incorporation of active flow control mechanisms, such as leading-edge flaps or vortex generators, can further enhance performance, even under varying wind conditions. These improvements directly translate into increased energy output and a better return on investment.

The development of improved energy storage solutions is also critical. Wind power, being intermittent by nature, needs effective storage methods to ensure a consistent energy supply. Currently, pumped hydro storage is a well-established technology, but its geographical limitations restrict its widespread implementation. Innovative technologies, including advanced battery systems, compressed air energy storage, and thermal energy storage, are being actively investigated. Research focuses on increasing storage capacity, reducing costs, and improving the efficiency and lifespan of these storage solutions. Their integration with wind farms will significantly improve the reliability and grid stability of wind energy.

Smart grid integration, as discussed earlier, is key to effectively managing the intermittency of wind power. However, further advances are necessary to optimize power flow, improve grid stability, and minimize energy losses. Advanced control algorithms, incorporating real-time data analysis and predictive modeling, can anticipate and respond to fluctuations in wind power output. This prevents grid instability, reducing the reliance on conventional backup power sources and maximizing the integration of renewable energy. Moreover, the development of sophisticated forecasting techniques, leveraging advanced meteorological data and machine learning algorithms, allows for more accurate predictions of wind power generation, enabling better grid management and energy dispatch.

Hybrid energy systems combine wind power with other renewable energy sources, such as solar power or hydropower, to enhance reliability and reduce reliance on fossil fuels. Such systems offer a synergistic approach, leveraging the complementary nature of different renewable sources to create a more consistent energy supply. The integration of these systems also allows for optimized energy management, maximizing the use of available resources and minimizing overall energy costs. Further research into effective strategies for integrating various renewable energy sources, including both on-grid and off-grid applications, will contribute to the widespread adoption of such hybrid systems.

Finally, the ongoing development of smaller-scale wind turbines, particularly for residential and microgrid applications, holds significant promise. These turbines provide an opportunity for decentralized energy production, reducing reliance on large centralized power plants and improving energy independence at a local level. Technological advancements are reducing the cost and enhancing the efficiency of smaller turbines, making them increasingly attractive to consumers and businesses. Their deployment promotes sustainable development, particularly in remote or underserved areas with limited access to the main grid. This approach will increase the overall societal benefit of wind energy. The advancements in these areas are crucial to unlocking the full potential of wind power as a sustainable and reliable energy source for the future.

www.ingramcontent.com/pod-product-compliance
Lightning Source LLC
LaVergne TN
LVHW022017200926
848481LV00007B/2441

* 9 7 9 8 3 4 8 5 8 8 5 4 0 *